S0-GQG-266

The *Mechanix Illustrated* Guide to

HOW TO PATENT
AND MARKET YOUR
OWN INVENTION

The *Mechanix Illustrated* Guide to

HOW TO PATENT AND MARKET YOUR OWN INVENTION

Marvin Grosswirth

David McKay Company, Inc.

NEW YORK

ACKNOWLEDGMENTS

The author wishes to express his thanks and appreciation to: Mr. Isaac Fleischmann, of the U.S. Patent and Trademark Office; Dr. Karl F. Ross; and Dr. E. Allan Blair for their assistance, advice, and information, while retaining for himself full responsibility for any error in fact or judgment that may have inadvertently crept into the book.

658.1141
G878

10 9 8 7 6 5 4 3

Copyright © 1978 by Marvin Grosswirth and Fawcett Publications, Inc.

All rights reserved, including the right to reproduce this book, or parts thereof, in any form, except for the inclusion of brief quotations in a review.

MANUFACTURED IN THE UNITED STATES OF AMERICA

Library of Congress Cataloging in Publication Data

Grosswirth, Marvin, 1930-
    The Mechanix Illustrated guide to how to patent and market your own invention.

    1. Patents.   2. Inventions.   I. Title.
II. Title: How to patent and market your own invention.
T339.G76   658.1'141   78-7309
ISBN 0-679-51205-5

FOR RONA
. . . because she asked.

CONTENTS

CHAPTER

1

A Great Idea

Don't just stand there—invent!

That inspiring slogan is emblazoned over the entrance to the Public Search Room of the United States Patent and Trademark Office in Arlington, Virginia. Millions of Americans have already taken that advice and if current activity is any indication, millions more will continue to do so for many years. Some observers lament that over the years, fewer and fewer patents are being issued to independent inventors; over 70% of all patents now being issued are granted directly to inventors who work for large companies or institutions and who assign their patents to their employers. Sometimes the company to whom the patent is assigned is a small, one or two person business established by an independent inventor, but more often, the owners of new patents are the corporate giants. Companies like General Electric, DuPont, Eastman Kodak, and others whose names are familiar in the world of commerce and industry account for most of the significant inventions being patented today.

Does that mean there is no place for that mainstay of

traditional American resourcefulness, the rugged individu-
alist? Not at all.

For one thing, despite the fact that patents are assigned
to companies, inventions are usually the product of one or
two individuals who are either employed by the company
holding the patent or who, in the interests of efficient marketing and limited finances, have turned over the rights to
their inventions for monetary reward. Recognition that good
ideas come not from committees but from individuals is evident in the patent laws of the United States, which specify:
"Application for patent shall be made by the inventor, except as otherwise provided . . . When an invention is made
by two or more persons jointly, they shall apply for patent
jointly . . ." (The "otherwise provided" disclaimer covers
instances in which more than one person took part in developing an invention but one or more of the inventors either
could not be located or was not willing to participate in the
application. Instances in which a person is considered to be
"under legal incapacity," or dead are also covered.)

Inventors are often caricatured as somewhat fuzzy-
brained, absent-minded geniuses who sit on high stools in
basement laboratories, fiddling with test tubes or playing
with mathematical formulas while waiting for inspiration to
strike. Of course, nothing is further from the truth. On oc-
casion, a brilliant flash of inspiration strikes the brain of a
creative and inventive individual. But most of the time, an
invention, no matter how significant or silly, derives from
need.

When Joseph F. Glidden of De Kalb, Illinois, devised his
invention, he probably did not see himself as a scientific
or mechanical wizard. Nor, in all probability, did he expect his invention to literally change the geography and
economy of America. His invention, covered by Patent No.
157,124, issued on November 24, 1874, was labeled, simply,
"improvement in wire fences." Glidden's "improvement" is
more commonly known as barbed wire. Without it, the

American West, with its vast expanses of grazing land and with no apparent practical means for fencing off that land, is likely to have developed quite differently.

Similarly, when John Deere developed a plow that was made of steel instead of iron, he was probably thinking of the farmers he knew whose lives would be a little easier because of his new construction. It was a simple concept, but it was one which undoubtedly had a major impact on the development of America's agriculture.

And it was because of need that Abraham Lincoln invented an inflatable device that could lift vessels over treacherous shoals. It was need that inspired Lillian Russell to patent a traveling trunk. It was need that caused Whitcomb L. Judson to find his place in history when, on August 29, 1893, he was granted Patent No. 504,038 for his invention of a product that is today quite literally as common as clothing: the zipper.

Certainly, some inventions were the result of arduous research. A group of scientists at Bell Telephone Laboratories were studying the characteristics of various semiconductors and discovered certain properties that suggested new areas of experimentation. The result was the transistor, an invention that revolutionized American technology. Transistors can be found in everything from nearly-invisible hearing aids to giant computers and communications satellites. Other inventions have also resulted from pure research and experimentation. For the most part, however, a good invention comes from a person with a need.

One of the most famous was an idea that occurred to Chester Carlson. Carlson was graduated from the California Institute of Technology and accepted a position as a researcher for Bell Laboratories. He soon discovered, however, that patent law was more to his liking and he took up his new career in Rochester, New York, where the headquarters of the Eastman Kodak Company is located. Before long, Carlson became aware of the difficulties and expense of copying the documents required in his work and began to consider the possibilities of a simpler, less expensive method. He was aware that Kodak, among others, held patents for most photographic processes used to make copies, but his desire for an alternative method persisted. Because of that persistence, he developed a process that created a static electrical charge on paper so that carbon or ink particles would be attracted to that charge. If that electrical charge could be placed on the paper in a particular pattern, it would be, in effect, another form of printing. With a little bit of money and a great deal of perseverance, Chester Carlson worked on his idea and eventually produced what is today known as the Xerox copier. At the same time, he established Xerox, Inc. The rest is commercial history.

R. B. Crean, a scientist with the technical service laboratories of Mobil Oil Corporation in Princeton, New Jersey, demonstrated in a personal letter how need produced an invention that earned him a patent:

There is a chemical that dissolves hydrogen sulfide gas (H_2S) out of certain streams to be found in petroleum refineries. H_2S not only "stinks"; it eats the bejabbers out of pipes and such. This chemical does the job just so long and then it has to be regenerated because it loses its power and is expensive to replace. Well, one scheme to regenerate the stuff: diethanol amine— "DEA"—involves a process involving electrolysis and uses cells and diaphragms and electrodes and such. It works, but, in the process, acids are formed on the waste side of the diaphragm at a rate that makes the process ten times more costly than the DEA we're trying to recover because it eats up the diaphragms. Here's where I come in: I use *two* diaphragms and pump a non-metallic base solution between the two to neutralize the acid as fast as it is formed. The diaphragms last indefinitely and the process works: U.S. Patent No.

2,983,656. This is an invention resulting from years of hard work.

The difference between the need of the petroleum processing industry and the average homemaker is primarily one of complexity. The homemaker, while not likely to be concerned with diaphragm-eating acids, is nevertheless faced with a variety of problems and frustrations. Many of these result in inventions so basic and simple in concept that one can only wonder why it took so long for someone to get around to them.

At an exposition of inventors held in New York in 1977, scores of such inventions were exhibited. A mother who got tired of her child's skin getting caught in zippers devised an "infant's neck protector." The creative mind of Theophilus Gibbings came up with: an automobile jack that can be operated from inside the car, a light-dimming switch that works in coordination with a timer, a portable illuminated makeup mirror, and a windshield blower that "is engineered to save time—while motorist warms engine, windshield blower cleans snow, fog or rain from shield. Most important is that the user no longer has to stand in the freezing cold, dusting snow from the windshield." Rose S. Ferrara of Leonia, New Jersey, developed and patented a simple basket-like device that hooks onto a mail slot so she no longer has to stoop down to pick up her mail. Ms. Ferrara also invented "Easy-On Bands for Rolled Meats . . . These thread-covered bands, which come in various sizes, are stretched and placed over the rolled meat . . . in place of conventional fastening means. These bands may be rinsed or washed and reused." Among other devices exhibited at this inventors' show were a wig shaker, a pet food dispenser, a shopping cart that could be hauled up stairs, a self-closing toilet seat, a guide for a saber saw, a telephone timer, and enough dog toilets to gladden the heart of any urban pedestrian.

What is remarkable about most of these inventions is not

their complexity but their simplicity. Surely other people must have thought it would be useful to have some device that would catch mail placed in a mail slot so the recipient would not have to stoop down to retrieve letters and catalogs, which would be otherwise strewn all over the floor. What makes Ms. Ferrara's invention different is that she had the good sense, first, to develop it carefully, and second, to apply for a patent.

Sometimes the need is created by the invention. Given the basic efficiency of matches, there is really no genuine need for cigarette lighters. But once the cigarette lighter *was* invented, it was promoted as a convenience; thus, although the "need" exists, it is a somewhat artificial one. Variations on the theme, such as disposable cigarette lighters, are also, essentially, artificially created "needs."

Sometimes the only "need" motivating an invention is the need to keep up with or surpass competition. Many of the single-function electrical appliances now being marketed—hamburger makers, doughnut makers—are not unlike the disposable cigarette lighter: first the concept of convenience is established and then the desire for that convenience creates the "need."

The ethics, morality, and economics of such created or technological needs are perhaps best left to those who have the time, talent, and money to ponder such matters. We will concern ourselves with the individual inventor who has an idea—a good idea, a workable idea, a practical idea, a patentable idea. If necessity is, in fact, the mother of invention, then perseverence surely is its father. It is a long, uphill road from bright idea to profits and rewards. But the road is by no means an impassable one and your great idea is the first step.

Don't just stand there—invent!

CHAPTER

2

What is a Patent?

In its simplest form, a patent is a grant by a government giving an inventor a monopoly on his invention for a fixed period of time. In the language of the U.S. patent law and the grant itself, a patent is "the right to exclude others from making, using or selling" the invention the patent covers. Thus, the patent does not tell the inventor what he must or may do; it gives him the right to prevent others from doing anything with his invention except as he specifies. Once an inventor obtains the patent, he may choose to bury it or do nothing further with it.

The word "patent" was derived from the Latin word meaning "open." Originally, *literae patentes*, or "letters patent," were issued by sovereigns to grant some special privilege, such as an exclusive dealership or a special right of passage.

The concept of letters patent to grant exclusive rights for inventions started four centuries ago. In 1594, Galileo obtained a patent for a system of land irrigation. He suggested that if his invention were protected by such a patent, he would be encouraged to develop other useful devices. That was, in fact, one of the basic underlying justifications for patents. In 1559, Giacopo Acontio, an Italian, appealed to Queen Elizabeth I of England for a patent on furnaces and "wheel machines" that he had invented. The Queen granted the request and is believed thereby to have inaugurated the British patent system, on which the American system is largely based. But it was not until the founding of the American republic that patents passed from common law to written legislation. Although the United States Constitution provided for the protection of inventors through patents, the first actual patent bill was not signed by George Washington until April 10, 1790. For the first time in history, the intrinsic right of an inventor to profit from his invention was recognized by law. Previously, it was the prerogative of a monarch or a special act of legislature that determined the privileges granted an inventor. With the enactment of the patent bill, many of the inventor's privileges became rights.

The Patent Act of 1836 effected some significant changes in the original law and is the foundation for the existing system of granting patents. The Act of 1836 reestablished the "examination" system that had been in effect prior to 1793. Once again, it became necessary to determine the novelty and usefulness of a patent application. The Act also reaffirmed the requirement that the determination of whether a patent is granted should be based upon "prior art." "Prior art" is a phrase well worth remembering. It refers to "that which has been invented or used before." The expression is also commonly used when discussing patentability (more about that later). Under the Act of 1836, the patent office became a separate bureau, with its own head, known as the Commissioner of Patents.

A number of changes were made over the years to refine the patent law, and the patent office shifted not only location but accountability, being at various times part of the Department of State. Today it is known as the United States Patent and Trademark Office and is under the jurisdiction of the United States Department of Commerce.

In a nation that prides itself on free enterprise and open competition, why should the government interfere in open competition by granting patents and all the rights and privileges that come with a patent? There are several good reasons, all of which serve to benefit the economy, the system of free enterprise, and the public.

First, by offering the protection of a patent, the government encourages invention. Without that protection, many an inventor would be reluctant to divulge his brainchild. Then the inventor would be deprived of any benefits that might accrue to him as a result of marketing his invention, the public would be deprived of the presumed benefits it would receive from the promotion of that invention, and industry would be deprived because many innovations are built on a foundation of "prior art." The protection of a patent offers the inventor an opportunity to refine and develop his product, to work out its production, to get it into the market place, to recover his investment and, if he is lucky, to realize some profit. It seems a fair enough proposition: in exchange for making your invention public and making it available to the people, you are granted the exclusive right, for a limited period of time, to the exclusive use of that invention.

The practical value of the philosophy underlying the concept of patents can be seen in the impact that a number of patents have had on the shape of the world. Had inventions not been protected, Eli Whitney might not have invented the cotton gin in 1794, a machine crucial to the development of the great American textile industry.

Also important was Cyrus H. McCormick, who in 1834 changed America's agriculture, as well as that of the entire world, forever by inventing the McCormick Reaper. Two years later, Samuel Colt patented a "revolving gun," the first of the famous six-shooters, which played such an important role in the winning of the West (not to mention the American film industry). In 1840, Samuel F. B. Morse patented the telegraph; in 1846, Elias Howe, Jr., patented an "improvement in sewing machines"; in 1873, Eli H. Janney patented a car coupler, which gave birth to the gigantic railroad industry of the twentieth century; in 1890, Ottmar Mergenthaler patented his linotype machine, changing the nature and cost of printing; in 1897, Guglielmo Marconi patented wireless telegraphy; in 1906, Wilbur and Orville Wright received Patent No. 821,393, for certain "new and useful improvements in flying machines"; and on and on and on—transistors, computers, plastics, fabrics, communications, aerospace, energy—but not all energy. Under the Atomic Energy Act of 1954, patenting inventions in the production of atomic energy weapons became specifically prohibited.

To be sure, not all inventions produced in the 200 years since passage of the first American patent law were earth-shattering. Along with John Deere's plow, Charles Goodyear's vulcanized rubber, Whitcomb L. Judson's zipper, and Thomas Edison's light bulb, there were patents for an automatic hat tipper, a dimple maker, a bustle with a hidden, built-in seat, a toilet seat designed to keep people from standing on it, a tapeworm trap, a bisexual bicycle seat, and hundreds upon hundreds of other inventions, all of which gave rise to the speculation that when the patent office said "useful," perhaps what they really meant was "usable."

One of the most interesting developments in the history of patents and the concept of invention occurred on October 6, 1977. The United States Court of Customs and Patent Appeals ruled that the Upjohn Company, a major manufacturer of pharmaceutical products, should be granted a patent for the microorganism known as *streptomices vellosus*, used by Upjohn in producing the antibiotic Lincomycin. This was the first time in history that a patent was allowed on a non-plant life form. "Microorganisms have come to be important tools in the chemical industry," the court said, "especially the pharmaceutical branch thereof. And when a new and useful tangible industrial tool is invented . . . we do not see any reason to deprive it or its creator or owner of the protection and advantages of the patent system." Although there are such things as plant patents, which will be discussed in more detail shortly, this new ruling possibly opened the door for a whole new series of patents on new forms of animal life.

Patent and Trademark Office

The primary function of the Patent and Trademark Office is the administration of patent laws as they relate to granting patents for inventions. Obviously, this includes the actual granting of the patents themselves. The PTO examines patent applications to determine whether the inventor is entitled to a patent under the law. The PTO publishes the patents that have been granted; it also issues various publications about patents and patent laws, records the assignment of patents, and maintains a Search Room for public use so that patents and related records can be examined. The PTO supplies copies of records and other papers when requested.

The head of the PTO is the Commissioner of Patents and Trademarks. His large staff includes several Assistant Commissioners, one of whom is in charge of the Patent Examining Group.

Examining patent applications is the PTO's most important function. The applications are divided among a number of examining groups, each of which has jurisdiction over certain assigned fields of invention. At the head of each group is a group director. The examiners study each application and determine whether a patent can be granted. If an examiner refuses a patent, his decision may be appealed to the Board of Appeals. The Commissioner of Patents and Trademarks may be petitioned to review other matters regarding a patent application. The examiners also determine when an "interference" exists among pending applications or between a pending application and a patent, in which case an examiner may institute interference proceedings. (Interferences, too, will be explained shortly.)

In addition to the examining groups, there are a number of sections, divisions, and branches of the PTO that perform various functions. According to the latest published figures (December 1976), the PTO has about 2,700 employees; about half are examiners and "others with technical and legal training." The PTO receives over 90,000 patent applications a year. These applications cover 3 types of patents specified by the patent law: *utility patents*, *design patents*, and *plant patents*.

A *utility patent* covers a new invention. It is issued for a period of seventeen years. A *design patent* covers "any new, original and ornamental design for an article of manufacture," and is issued for a period of three-and-one-half, seven, or fourteen years, depending upon the term requested by the applicant. A *plant patent* covers "any distinct and new variety of plant, other than a tuber-propagated plant, which is asexually reproduced." Plant patents are also issued for seventeen years, (For purposes of this book, we are mostly discussing utility patents. Design patents are discussed in some detail in Chapter 8.)

Patent Myths

Over the years, some interesting but totally false myths have risen about patents. Perhaps the most popular is the one about a Commissioner of Patents who, at some vague time during the nineteenth century, is alleged to have resigned his post, in the belief that everything that *could* be invented had *been* invented. No one seems to know exactly how this strange tale originated. The best guess is that it was prompted by a statement by Henry L. Ellsworth, Commissioner of Patents, who, in his Annual Report for 1843, said:

"The advancement of the arts, from year to year, taxes our credulity and seems to presage the arrival of that period when human improvement must end."

Another popular myth is that an ordinary layman—an office worker, a factory hand, a farm worker, a homemaker—can invent something that for some reason the trained technological experts of science and industry have overlooked and become rich. If that has ever happened, no one seems to have documented it with facts and figures.

It is perhaps the American cartoonists who gave rise to the myth that every patent application must be accompanied by a working model of the invention. In fact, there was a time when models were required by the PTO, but that requirement was discontinued in 1870. The PTO still has the right to demand a model to prove that the invention is operable, but requests for such models are rare.

There is a widely held belief that a U.S. patent protects an invention in foreign countries. As a matter of fact, however, if you want your invention protected in a foreign country, you must obtain a patent in that country. In general, the reverse is also true: an invention patented abroad is not protected in the United States by the foreign patent. You need a U.S. patent to protect your invention in this country. How to determine whether your invention is patentable is the subject of the next chapter.

CHAPTER

3

Developing Your Invention

According to the patent law, anyone who "invents or discovers any new or useful process, machine, manufacture or composition of matter, or any new and useful improvements thereof, may obtain a patent"—within, of course, the restrictions of the law. You could not, for example, obtain a patent for an invention whose end result would be illegal. Thus, you could patent a device that detects a "bug" on your telephone, but not one that prevents the recording of telephone calls to avoid being charged for them.

Let's take a look at some of the words in that quotation from the patent law. *Process* refers to a method or means of arriving at some end result or end product in a wholly new way, even though that end result may not in itself be new. For instance, there is nothing particularly new or unusual in the physical properties of the book you are now holding. However, although the basic design and composition of books are probably older than printing itself, it is conceivable that at this very moment patents are being applied for that cover new ways to make paper, new methods of setting type, innovative processes for bookbinding, etc. The word *machine* means exactly what it says; a machine is a mechanical device that performs some sort of function, whether it is sorting coins at a local bank or flying people halfway around the world at the speed of sound. The word *manufacture* covers items that are made, whether laboriously put together by hand in a basement workshop or ingeniously assembled by complicated machines. *Composition of matter* refers to chemical compositions. It includes not only entirely new chemical compounds but, under certain circumstances, may include mixtures of ingredients that are unique and innovative. According to the United States Patent and Trademark Office (PTO), these classes of subject matter, taken together—*process, machine, manufacture,* and *composition of matter*—include practically "everything which is made by man and the processes for making them." Some nit-pickers may want to question that conclusion, but they will be hard pressed to find any exceptions.

The patent law also specifies that to qualify as patentable, an item must be "useful." Usefulness is a quality open to considerable discussion. An inventor may develop a simple faucet attachment so drinking water can be had in a variety of colors without affecting its taste or purity. Just how "useful" such an invention might be, however, may lead to some interesting correspondence between the inventor and the PTO.

In addition to being useful, the subject of a patent application must also have what the PTO calls "operativeness." It is not enough, in other words, to demonstrate that the intended purpose of a machine is a useful one; you must show that the device operates in such a way that it achieves that result. The PTO states flatly: "Alleged inventions of perpetual motion machines are refused patents."

Perhaps the most important condition for patentability is novelty. The statute covering patents states that "an application for a patent will be denied" if:

(a) The patent was known or used by others in this country or patented or described in a printed publication in this or a foreign country before the invention thereof by the applicant for patent, or

(b) The invention was patented or described in a printed publication in this or a foreign country or in public use or on sale in this country more than one year prior to the date of the application for patent in the United States . . .

Thus, if someone has a bright idea for a new process or technique and publishes an article about his inspiration in a magazine or journal, that publication may cost him the right to patent his idea. It doesn't matter how obscure the publication is, in what language it is printed, or how limited its circulation may be. According to the law, publication and patents are not compatible. Even telling people about the invention might cancel its eligibility for a patent. There are some fine points, as evidenced by the apparent contradiction of paragraphs (a) and (b) quoted above. Actually, the differences have to do with the actual making of an invention. Under certain circumstances, if the inventor has already created the object of his inspiration, he may be able to obtain a patent anyway, provided no more than a year has gone by since publication or sale of the item. For the novice, however, it is best to remember that the only way to protect an invention and obtain a patent for it is to keep quiet about it.

Surely it is not necessary—but we will do so anyway—to mention that a patent will not be granted on anything already invented by someone else. Patent attorneys all have stories about people who come to them with sensational, money-making ideas, which originated not in the would-be client's imagination but on a trip to a foreign country or in an old magazine article. One patent developer told of a woman who telephoned his office every day for a week with two or three interesting and useful ideas for household tools and appliances. Finally, suspecting that the source of these ideas was something other than the woman's fertile brain, he asked her where she was getting her inspirations. Quite innocently, she informed him that her reference source was the current Sears, Roebuck catalog.

There are five broad rules that should be applied to determine whether an invention is patentable:

1. It must be original and novel.
2. It must be unknown.
3. It must be useful.
4. It must be workable.
5. It must be beyond the idea stage.

That last rule requires some explanation. Stated flatly, an idea may not be patented. In a sense, there is some redundancy in that observation. It follows that if one of the conditions for a patent is operativeness, the possibility of patenting an idea by itself is automatically eliminated; you must show that the idea is operative. Still, it is worth remembering that it is not enough to come up with a good idea. As far as the PTO is concerned, you must demonstrate that the idea can be put into effect in a practical way.

Even if the five basic rules are met, there is still some question as to what is and is not patentable. For example, a particular function, such as paper making or printing, is probably not patentable, but the method by which that function is achieved could be. On the other hand, a function that is completely new might be patentable. Similarly, you will probably be refused a patent if you design a machine part that simply replaces an existing part, or if the material from which the part is made is different. But if you can develop a machine that achieves a known function but has fewer working parts, your chances of being granted a patent are excellent.

Our correspondent, R. B. Crean (the scientist who works for Mobil Oil Corporation and whom we met in Chapter 1), offers an example of an invention, which, in his opinion, should have been granted a patent but was not:

> I invented a device that will pump the gunk, sludge, etc., out of underground or otherwise inaccessible storage tanks for gasoline, fuel oil, etc. There are a few patented devices designed to do this job but they plug up with a frequency that renders them commercially unfeasible. My design is such that plugging is impossible—and people who are involved in the business of, say, retailing home fuel oil, are making and using these devices with success . . . This "invention" was denied by the U.S. Patent and Trademark Office . . .

Some Points to Consider

In developing your own invention, there are some important points to consider besides the five basic rules outlined above. First, make sure that your invention is a *practical* one. One of the first determinations is whether there is a genuine need, beyond your own, for such an invention. You may be an amateur photographer who happens to be seven feet tall and who therefore feels a pressing need for a very strong but light tripod, which is also very long and collapses to four inches when not in use. But you must ask yourself how many other seven-foot-tall amateur photographers there are in the world who would be willing to pay for such a tripod. Even if you determine that there is a need, you must ask yourself whether existing products already fulfill that need. There may be tripods that are not as tall, as light, as compact, or as sturdy, but which nevertheless adequately meet the requirements of other seven-foot-tall amateur photographers.

The second most important factor to consider is cost. The cost of developing an idea and ultimately obtaining a patent is impossible to estimate in a general discussion like this. (There are, of course, specific fees involved; these will be discussed later.) Understandably, development costs vary with the particular invention. But even beyond that, cost is an important consideration: the would-be inventor must determine what it will cost to manufacture and market his product. Your tripod for seven-foot-tall photographers may be practical, it may fulfill an existing need, and may have a considerable potential market, but if the final retail price is too high, no one will buy it.

Now, let's look at some of the other factors you should consider when determining whether to turn your idea into a physical reality.

Compatibility. Is your product a one-shot, one-function device or does it have several uses and applications? That tripod we have been talking about will have greater usefulness—and, therefore, greater marketability—if it will accept all types of cameras, if it can be readily converted to a light stand, if it can be used as a support for backgrounds and if, with a simple accessory, it converts to a seat.

Ease of operation. Unless you are planning a process or product for some complex technological or scientific purpose, your invention should be relatively easy to use. If our hypothetical tripod comes in several pieces that must be screwed or bolted together by the user, or if it requires the manipulation of a series of locks and catches, it will have little market appeal. Similarly, it must be safe to use. Sharp, unprotected points, places where fingers might get caught, and protrusions capable of putting out an eye will greatly reduce the appeal of the product.

Efficiency. The invention, to be practical, must smoothly and efficiently perform the function for which it is intended. Our tripod is destined for the rubbish heap if one of its legs has a tendency to slip during time exposures.

Marketing factors. The practicality of actually selling the product is an extremely important consideration, which we will discuss at some length later on. However, when determining at the outset whether an idea is worthy of further development, it is important to think about the practical aspects of selling the product. If it is apparent that most seven-foot-tall photographers live in one area (for example, the southwestern part of the country), then it may be possible to establish a distribution program that is relatively uncomplicated and inexpensive. If, on the other hand, they turn out to be scattered all over the country, they represent an extremely limited market for individual retailers. The product may not lend itself to such alternate selling channels as mail order.

Timing is another important marketing factor to consider. The person who designed and began marketing a desk-top stand for hand-held calculators must have had a superb sense of timing as far as marketing is concerned. However, there may be, among the multitudes of unknown inventors, someone who devised a very clever but totally unmarketable rack for hula hoops.

Timing involves not only the product's present and immediate future market, but also the possible long-term developments. Again using our tripod as an example, the inventor might take into account the fact that 35mm cameras are becoming smaller, lighter, and more compact each year. He might also consider that, to a lesser extent, the same is true of motion picture cameras. Then, to be practical, he might develop the idea of the tripod, on which a photographer could mount and use both a 35mm still camera and an 8mm movie camera simultaneously.

Other manufacturing trends and technological developments should also be considered. Clearly, the tendency today is to move from metals to plastics whenever possible. Bearing in mind the possibility that a plastic material rigid enough and sturdy enough may emerge in the near future, the inventor of our tripod may want to develop the product so it can be easily converted from metal to plastic components, when and if that sturdier plastic material becomes a reality.

Manufacturing. If the new invention can be manufactured in already existing facilities with little or no modification, it stands a far better chance of success than if it requires expensive tooling or completely new operations. It may be possible to manufacture our tripod with a few simple adjustments of machinery already in use. Sometimes, a modification or two in the invention itself eliminates the need for wholly innovative processes. The more expensive it is to set up the manufacturing of an item, the less likely it is that the item will ever be manufactured.

Even if the setting-up costs can be met, serious consideration must be given to the manufacturing costs. Raw or semi-processed materials should be readily available and at a price that makes the finished product accessible to the product's likeliest consumers. Careful attention should be given to minimizing the quantity of raw materials that goes into an invention and to reducing, as far as possible, the labor costs involved in assembling the product.

Legal considerations. The law works both for and against inventors. New regulations on safety, environmental protection, energy conservation, and the like offer vast opportunities for developing new devices. But the law also stops an inventor whose idea may be ingenious but also either antisocial or in direct violation of some law. If one leg of our hypothetical tripod converts, with a flick of the wrist, to a sword or a bayonet, a patent is likely to be denied on the grounds that specific laws exist against carrying concealed weapons. Similarly, if the tripod is designed to readily adapt for use in distilling whiskey, the PTO is certain to frown on the patent application.

Stop for a moment and briefly review all the recommended requirements for determining the practicality of developing your idea into a patentable invention. At first glance, the list seems formidable and full of negatives. On second thought, however, you are likely to discover that your good idea meets these criteria. If it falls short on a couple of points, very often a simple change in one or two of the characteristics of your invention is enough to bring the product into line. Size, shape, color, material, design—any of these is an area in which slight modifications are possible. If we stay with our hypothetical tripod, we can easily see how some manufacturing problem might be reduced or eliminated if the inventor determines that a tripod, which collapses to five or six inches in length, is just as marketable as one that collapses to four inches. Other problems quickly dissolve if the inventor originally thought of making his tripod out of wood or steel but is now giving serious consideration to aluminum. Even if it turns out that there are not enough seven-foot-tall amateur photographers to constitute a market for the product, if the tripod is truly unique or can be manufactured by a previously unknown process, the product may find a market among *all* amateur photographers. Thus, if your idea fails to conform to the basic considerations discussed above, that does not mean you should forget about inventing, patenting, and marketing it. It does mean that you should think more about your idea to determine whether, in a practical way, it can be brought into reasonable conformity with the guidelines.

Developing the Invention

As was stated earlier, it is not enough to have an idea; it is necessary to develop that idea into working reality. As you proceed with the development of your idea, it is absolutely essential that careful records be kept and that they be kept in a particular way. Sometimes disputes arise over who is the original inventor of a particular invention. When making their determinations of originality, the PTO and the courts consider two major factors: *conception* and *reduction to practice.*

Date of conception refers to the exact moment when the idea for the invention occurred to you. *Reduction to practice* means that the invention has been worked on and developed by you to the point at which it has been successfully tested. Should a patent application be rejected on the grounds that the invention is not practical or operative, proof that the inventor has reduced the invention to practice may be necessary.

A number of factors influence the determination of what constitutes diligent reduction to practice. It may be possible to produce a working model of a tripod design for seven-foot-tall photographers in a basement workshop or, at some expense, in a commercial machine shop. A prototype for a new locomotive or airplane may be somewhat more difficult to turn out. In any case, careful records are absolutely essential, not only because you will need them if some dispute arises over who got what idea first, but because accurate and detailed records are the only way you will be able to chart the development of your idea into a practical, workable invention. There are some rules to follow that will greatly minimize the difficulties in your quest for a patent:

1. *Don't talk.* You already know that once an invention is published, it is no longer patentable. Literally, the word *publish* means "to make public." In effect, therefore, if you discuss your idea as part of an after-dinner conversation or in some other casual setting, you may lose your protection for the idea. Still, it may be necessary when determining the practicality of your invention to discuss it with some experts. Should that become necessary, then the first person to talk to is your lawyer. He will be able to put you in touch with people whose integrity and honesty can be relied on.

2. *Keep complete and detailed records.* We'll discuss the specifics of record keeping below.
3. *Keep your idea alive.* Continue to conduct experiments, make tests, and do whatever is necessary to show that you have exercised a reasonable amount of diligence in "reducing to practice" from brilliant idea to working invention.

Record Keeping

Now let's discuss the record keeping, which is so vital in obtaining and holding onto a patent. There is an important exception to the injunction against discussing your invention with others: find two friends whom you know well and whom you can trust to act as witnesses throughout the reduction-to-practice process. The witnesses must not be related to you. Begin your record keeping with a bound notebook, one in which pages cannot be inserted or removed easily. Enter a description of the invention and include sketches, diagrams, and drawings. Have the description and drawings signed and dated by the two friends acting as witnesses, indicating that they have understood what they have read. Have the witnessing notarized.

Make sure each page is numbered consecutively. Do not leave any blank lines or spaces. Draw a diagonal line through any blank areas. To completely eliminate any future doubt, try to have each page dated and witnessed. It is unnecessary for the witnesses to sign the pages on the same day they are completed, so long as the signing is done within a reasonable time—a couple of weeks, or a month at the most.

As you continue developing your invention, have the witnesses sign the pages, each time indicating that they clearly understand what your invention is, what it is supposed to do, and how it works.

When keeping records of your work, excessive neatness

could be a definite liability. Make all your entries in ink and *never erase anything*. If you make a mistake, arrive at a wrong conclusion, or turn in a direction that proves to lead nowhere, neatly cross out the incorrect data; however, make sure it can still be read. You can offer no better proof that you have been diligently working at reducing your invention to practice. When you cross out the incorrect data, initial and date the crossing-out.

Do not withhold any information from the record. Even if you have second thoughts about the integrity of your witnesses, disclose everything nevertheless; the likelihood of a witness stealing your invention is slim. Even if that should happen, the record book is proof of prior invention. Furthermore, the scoundrel will have a difficult time proving he did not steal the idea if his notarized signature shows up in your record book. On the other hand, missing details may make it impossible for a witness to state that he fully understands what your invention is all about.

Do not throw away papers pertaining to your invention. Hang on to sales slips for purchases connected with the invention, including the record books themselves. Retain any sketches, notes, or diagrams you may have made while working in a shop or laboratory. Carefully preserve any correspondence you may have had with anyone about the invention.

It is a popular misconception that you can protect your invention by carefully describing it in writing, inserting that writing into an envelope, and sending the envelope to yourself by registered mail. It is assumed that if the envelope, with the dated post office seal over the flap, is unopened, proof of the conception date is established. Unfortunately, such a registered letter is completely useless. "Your priority right against anyone else who makes the same invention independently," advises the PTO, "cannot be sustained except by testimony of someone else who corroborates your own testimony as to all important facts, such as conception of the invention, diligence, and the success of any tests you may have made." The importance, therefore, of carefully maintained records, frequently witnessed, with signatures occasionally notarized, cannot be over emphasized.

Now that you have determined the practicality of your invention and have reduced it to practice, you are probably ready to think about obtaining a patent.

CHAPTER

4

Getting the Patent

Registered Practitioners

It is possible for an individual inventor to apply for and obtain a patent without outside help. Any number of people have done so. But it is the consensus of all the experts, including those at PTO, that the task will be quicker, easier, and, in the long run, probably cheaper if you engage the services of a *registered practitioner*. There are two types of practitioners: *patent attorneys* and *patent agents*. Both are required to register with the PTO. Agents are individuals permitted to "practice" before the PTO, which means they can represent you in all PTO proceedings. They are knowledgeable in the details of preparing applications, including language style, drawing style, and claims. A patent attorney can do everything an agent does and has the additional advantage of being able to represent his client in court if necessary.

WARNING: Registered practitioners—patent attorneys and patent agents—are specifically prohibited by the PTO from advertising. They are permitted to list their names and addresses in certain directories, such as a classified telephone directory, but if you see an advertisement in a newspaper, a magazine, or in the mail for a firm or individual offering to help you with any or all phases of obtaining your patent, the mere existence of the advertisement tells you the advertiser is not a registered practitioner. The Patent and Trademark Office will send you a list of registered practitioners in your area. You can also examine the *Directory of Registered Patent Attorneys and Agents* at the nearest field office of the U.S. Department of Commerce or at the public library.

The Search

The first step in deciding whether you have a patentable invention is to make a search and determine whether someone has gotten there before you. Ultimately, your search may lead to that vast storehouse of technology known as the Public Search Room at the Patent and Trademark Office in Arlington, Virginia. But before that happens, and even before you go through the expense of engaging the services of a registered practitioner, you may want to make a preliminary search on your own. You can probably do that without wandering very far from home.

Briefly, a patent search is an examination of the patents and publications relating to your particular invention. Remember our tripod for seven-foot-tall photographers? To determine whether we truly have a unique invention, we must examine previously issued patents covering tripods and, perhaps, similar devices, such as adjustable drapery hardware, collapsible blackboard pointers, self-adjusting furniture legs, and any other device that might incorporate some of the characteristics of our tripod. The search will show "the state of the art" on tripods. We will learn whether there is, in fact, anything new under the sun about our invention or whether "prior art" is such that patentability of our tripod is in doubt. We may even learn about improvements that could be incorporated into our own invention.

To begin your patent search close to home, check with a nearby public or university library; over 400 of them receive the *Official Gazette* or the *Index of Patents*. The *Official Gazette* is a thick volume that has been published by the PTO every week since January 1872. Each volume contains an abstract and one drawing from every patent issued that week. It also includes notices of patent and trademark suits, indexes of patents and patentees, a list of patents available for license or sale, and other information of a more general nature, such as changes in classification and changes in rules. The *Index of Patents*, also published by the PTO, is the annual index to the *Official Gazette*. It consists of two volumes, an index of patentees and an index of patents, arranged more or less by category.

You may come across a patent in the *Official Gazette* that you feel deserves a closer look. If so, you may be able to find the patent in one of the more than twenty-five patent depositories serving major metropolitian areas (see Appendix B). These depositories, located in public and university libraries, have copies of patents on file, which may be examined by the public. Your search can be made somewhat easier if you write to the PTO and ask for the classes and subclasses of patents most likely to cover the invention in which you are interested. If any patents are of such interest that they require further detailed study, you can order copies of the patents from the PTO at 50¢ per sheet ($1.00 minimum charge per order). The mailing address of the Patent and Trademark Office is: Washington, D.C. 20231.

Even if this preliminary search does not turn up any information indicating that your invention is not patentable, you are not yet home free. It is now necessary to make a complete and thorough search of all existing patents. That means a trip to the Public Search Room at the Patent and Trademark Office. This office is located in Crystal Plaza, a modern office complex just off Jefferson Davis Highway in Arlington, Virginia, a short trip by bus or taxi from downtown Washington. Whether you make that trip yourself or engage the services of a professional searcher is a decision you will have to make yourself. Often, however, the cost of the search done by a professional is less than the cost of a trip to Washington. This is also the time to decide whether you want to proceed on your own or whether you should seek out a registered practitioner—a patent attorney or a patent agent.

The Public Search Room is a cavernous, two-story library of technology. Patents are arranged according to classes and subclasses. At the entrance to the Public Search Room is a friendly receptionist who will guide you to a huge volume called "The Manual of Classification." In this manual, you will find the classes and subclasses most likely to apply to your invention. There are more than 300 main classes and over 90,000 subclasses into which the patents are divided. In late 1976, the PTO happily announced the issuance of Patent No. 4,000,000 (four million). When you consider the number of patents in existence, the number of main classes and the number of subclasses, it becomes apparent that a patent search can be a formidable job. Patents issued since July 10, 1962 are recorded on microfilm in numerical order and may be examined in a microfilm reader. But the patents begining with No. 1, issued in 1836, through those issued up to July 10, 1962, are in bound volumes around the Public Search Room, and all but the most intrepid inventors soon begin asking themselves whether it would not be wiser to have their search done by a professional. If, therefore, you have reached the point at which you feel a search in the Public Search Room is warranted, this is the time to decide whether to hire a patent attorney or a patent agent, who can either conduct the search or provide a skilled searcher.

The Test of "Obviousness"

If you have determined that your invention is unique, operational, and useful, then the time has come at last to apply for a patent. But to qualify for a patent, your invention must be just that—an invention. One of the factors that could stand in the way of your being granted a patent is obviousness. If your invention is one that would be obvious to anyone having the usual skill or knowledge in a particular field, then it will not qualify for a patent. For example, think of all of the times you have used an ordinary paper clip or a wire coat hanger for some purpose other than the one for which it was originally intended. That would not mean you could obtain a patent for a slightly modified paper clip or coat hanger to be used for one of those other purposes. Similarly, a change in some basic characteristic of an object is not enough to obtain a patent. A paper clip remains a paper clip whether it is oblong, circular, or triangular; whether it is made of metal, plastic, or cardboard.

On the other hand, if such a simple change results in some spectacularly unusual effect, a patent may be issued. One example of this was a machine used to manufacture paper. For years, the paper manufacturing industry felt hampered because this otherwise efficient piece of equipment was slow. Eventually, someone discovered that by lifting one end of the machine about a foot from the floor, the machine's production rate was increased by 100%. Although the change was a relatively minor one, a patent was granted.

The Test of "Aggregation"

An aggregation occurs when two or more already existing objects are combined to form a new object that performs no new function. A friend of mine was recently enthused over such a device, which he called an indoor sundial. He envisioned the sundial mounted on a rotating mechanism, over which was suspended a light source. The light source was to be fixed and the rotation of the sundial was to be controlled so he could accurately read the time from the sundial. It was pointed out to him that what he had was an aggregation of a clock and a lamp and that, in all probability, the patent would be denied. Sensibly, he abandoned the idea.

Filing the Application

The application begins with a request to the Commissioner of Patents for the patent. This is called the *petition*. It includes your name, your address, and your citizenship. (It is not necessary to be an American citizen to receive a United States patent.) Also included in the petition are the name and address of your patent agent or attorney.

A properly prepared petition reads as follows:

To The Commissioner of Patents and Trademarks:

Your petitioner, ..., a citizen of the United States and a resident of ...., State of ...., whose post office address is ...., prays that letters patent may be granted to him for the improvement in ...., set forth in the following specification; and he hereby appoints ...., of ...., (Registration No. ....), his attorney (or agent) to prosecute this application and to transact all business in the Patent and Trademark Office connected therewith. (If no power of attorney is to be included in the application, omit the appointment of the attorney.)

Following the petition is an *abstract*, a short summary, in nontechnical language, of the invention. Ideally, the abstract is between fifty and one hundred words long.

Next comes the *Summary of the Invention.* The summary describes the general nature of the invention and may include information about existing products or prior patents that relate to the invention, explaining why the invention being applied for is superior. This is usually followed by several short paragraphs, which give the objectives of the invention, explain how they are implemented, and enumerate the reasons why these objectives have an advantage over the inventions or products already in existence.

Next is the *Reference to Drawings,* which gives a very short description of the accompanying drawings (e.g., "Figure 1 is a front view of my invention," "Figure 2 is a side view of my invention," "Figure 3 is a cross-sectional view of my invention").

The *Specification* follows. The Specification must be a complete and detailed description of the invention. It must explain how the invention is constructed, how it works, and how it is used. While the description must be complete, it must also be clear and concise. It must be stated with such exactness that anyone familiar with the field to which the invention applies could make the invention from the specifications.

As part of the careful description in the specification, it will probably be necessary to refer to accompanying drawings. This should be a step-by-step process, which begins with generalities and works its way down to specifics.

The specification serves two purposes. First, it describes the invention by referring to the numbered parts on the accompanying drawings. Second, it forms the basis for claims to be made for the invention. Anything not included in the Specification may not be included in the Claims section.

Claims

The *Claims of Novelty* follow the Specification. Claims tell the patent examiner what makes your invention different from all the other products now on the market, different from the inventions already covered by patents, and different from all the inventions described in publications and periodicals.

Writing a Claim is a tricky business, requiring creative thinking and knowledge of virtually everything there is to know about the object for which a patent is being sought. The inventor must be able to describe the abstract aspects of his invention with precision. If the inventor or his agent has been diligent in conducting the patent search, the inventor will know, when writing his claims, how to avoid running up against previously granted patents. In general, the broader and more simply stated a claim is, the stronger it is likely to be. If the inventor of our hypothetical tripod specifies in his claim where every rivet, nut, and bolt should be placed, his patent may prove to be of little use to him. In such cases, a competitor could probably place rivets, nuts, and bolts elsewhere without running afoul of the patent.

Generally, claims can be divided into two types: *generic* and *specific*. A generic claim encompasses all aspects of the invention. For example, "tripod" refers to all kinds of tripods. "Collapsible tripod" refers to a lesser number of tripods and a specific type. "Portable, collapsible tripod" is even more specific and more restrictive. Thus, in general, the longer the claim, the narrower it is likely to be. Therefore, if we are seeking a patent for our tripod, it is better to refer in the claim to "a tripod" than to one of the more detailed descriptions. Still, if there is a great deal of "prior art," it is necessary to narrow the claim to avoid or circumvent what already exists. If that necessitates an extremely detailed and extremely narrow claim, then obtaining the patent may prove to be more trouble than the invention is worth.

When preparing the claims, keep in mind these guidelines:

1. Remember that you are attempting to obtain a patent on the basis of the claims. You should, therefore, try to

restrict the claims to only those points that can be patented. For example, if you describe an element of the tripod as being made of metal tubing, you have probably eliminated an element of patentability. Metal tubing as such as not patentable. If, on the other hand, a particular method used in the manufacture of the tubing renders the tripod particularly rigid, particularly light, and particularly amenable to the type of construction you have in mind, then a description of some specific kind of tubing and an indication of the particular kind of metal is required. They are required because the functions the inventor has in mind are probably patentable.

2. Try to make sure that each element or component of your invention is covered by one or more claims. If, for example, our tripod consists of three major parts—the legs, the platform to which the legs are attached, and the pan head (the part on which the camera is mounted and which is usually manipulated horizontally and vertically by a handle)—there should be a claim for each part. If possible, each claim should try to include one or all of the other parts.

3. You should clearly delineate in the claim the physical or structural characteristics of the part it covers. You must show that there is a cooperative connection between the various features of the invention. It will not do to describe a pan head designed to sit on top of a tripod, and then describe a tripod designed to receive a pan head, expecting both elements to be accepted as claims in a single patent. You must show some interaction betweeen the pan head and the tripod.

4. You should avoid repetition in claims unless you modify your statement in the second claim by adding something to it. For example, if a claim describes a tripod leg as having seven collapsible sections, a subsequent claim could then include a tripod leg with seven collapsible sections, whose last is removable and reversible so as to

expose a sharp point, which can then be inserted into soil or sand.

5. You will find, if your search leads you to earlier patents, that the claims in those patents refer to the drawings by numerical references. This is no longer acceptable. Your claim must be descriptive within itself and without reference to the accompanying drawings.

6. You should avoid statements that have nothing to do with achieving the desired effect of the invention. For example, it is highly unlikely that the color of a tripod will in any way affect its efficiency, and there is no point in mentioning it unless it can be demonstrated that a given color is involved in its function.

Domination is the term for a patent claim containing every aspect of an invention. If an invention consists of six or seven parts, the claim including those six or seven parts is the one that dominates. It is important to include such a dominant claim since, without it, it may be possible to circumvent the patent.

Circumvention is the term used to get around an existing patent by, in effect, attacking that patent in its weak claims. For example, the patent for our tripod may cover only one "species"; that is, the inventor may have made claims only for a tripod suitable for seven-foot-tall photographers. We could have, however, included other species: tripods for medium-height or short photographers. If the patent claims refer specifically to the length of the tripod's legs or of the various sections of the legs, it may be possible to successfully circumvent that patent by using different lengths. Similarly, if the patent specifies the kind of metal to be used in producing the tripod, it might be circumvented by using another metal or plastic.

Drawings

Another important part of your patent application is the artwork. (Obviously, if you are patenting a chemical formula, drawings are not required. But in most cases, it will be necessary to provide them.) Drawings must be made on two- or three-ply Bristol board, 8½″ × 14″. They must be done in India ink and must maintain a 2″ margin at the top and ¼″ margins at the sides and bottom. This gives an overall drawing area of 8″ × 11¾″. Remember, to be considered fairly by a patent examiner, the drawings must be large and clear. When possible, different views—top, side, perspective, cross-sections—should be provided. All parts should be consecutively numbered. When one part appears in various views, that part must be identified by the same name and numeral in each view.

The PTO's established set of standards for drawings are fairly detailed and rigid (see Appendix A).

The Oath

Your patent application is completed by an oath, which is signed by you and notarized. A properly prepared oath looks like this:

................., the above-named petitioner, being sworn (or affirmed), deposes and says that he is a citizen of the United States and resident of ...., State of ...., that he verily believes himself to be the original, first and sole inventor of the improvement in .... described and claimed in the foregoing specification; that he does not know and does not believe that the same was ever known or used before his invention thereof, or patented or described in any

printed publication in any country before his invention
thereof, or more than one year prior to this application,
or in public use or on sale in the United States more than
one year prior to this application; that said invention has
not been patented or made the subject of an inventor's
certificate in any country foreign to the United States on
an application filed by him or his legal representatives
or assigns more than twelve months prior to this application; and that no application for patent or inventor's
certificate on said invention has been filed by him or his
representatives or assigns in any country foreign to the
United States, except as follows: ....

. . . .
(Inventor's full signature)

State of . . . .
    County of . . . .   ss:
    Sworn to and subscribed before me this . . . . . . day of
. . . ., 19. . . .

. . . .
(Signature of notary or officer)

. . . .
(Official character)

The Petition, the Specification, the Claims, and the Oath
must be in the English language and should be typewritten
on one side of the paper. The PTO prefers legal-size paper,
8-8½" × 10½-13", double spaced, with margins of 1½" on the
left-hand side and the top.

If by now you have the impression that filing a patent
application demands a considerable amount of attention to
tedious detail, you are entirely correct. But fulfilling that demand is not necessarily difficult. A degree of diligence, some

perseverance, a little intelligence, and some expertise are required. Of these, the expertise is the easiest to come by. By reading through several patents on items related to your invention, you will quickly become familiar with the style of both the language and the drawings, as well as with the exact format to be used. Another and highly recommended method of obtaining that expertise is to hire it. A patent agent or attorney is thoroughly familiar with preparing patent applications and invariably works closely with a reliable and trustworthy draftsman who can prepare acceptable drawings.

Charges and Fees

How much you pay the registered practioner and the draftsman for their services depends on a number of factors: the geographical location of the parties involved, the complexity of the claims and drawings, and the number of pages required to completely describe your invention.

Other fees, however, are fixed. The basic cost for filing an application for a patent (other than a design patent) is $65. For each claim in excess of one, there is a $10 charge, up to ten claims. For each claim over ten, the charge is $2. When the patent is issued, there is an issuing fee of $100, a $10 charge for each printed page of specifications, and a $2 charge for each sheet of drawings.

For design patents, the application filing fee is $20. On issuance, there is a $10 fee for three-and-one-half-year patents, a $20 fee for seven-year patents, and a $30 fee for fourteen-year patents.

The PTO has a fairly lengthy and detailed schedule of fees for other contingencies, such as filing claims independent of the original application, making certain appeals, and correcting an applicant's mistakes.

How much your patent will cost you, then, depends on

how well the application is prepared, how complicated the specifications, claims, and drawings are, and other factors. In addition, if you use their services, you must pay the cost of a registered practitioner, a professional searcher, and a professional draftsman. It is, therefore, difficult to estimate the cost of any one patent. In general, an uncomplicated patent, consisting of only a few claims and specifications and only one sheet of drawings, probably costs somewhere between $1,000 and $1,500, including the cost of professional services.

When facilities permit, the PTO provides certain services at—given the going market rate—very reasonable costs. For example, the PTO writes translations at a rate of $5 for every hundred words of the original language, makes patent drawings at a minimum cost of $25 per sheet, and makes corrections in drawings for a minimum charge of $3. It also mounts unmounted drawings and photoprints accompanying patent applications for $2 each.

"Prosecuting" Your Patent Application

When your properly prepared patent application is received at the PTO, it is turned over to an examiner who will make a complete search of all previous patents relating to your invention. The examiner will also search for existing literature on products relating to your invention. The first question that naturally arises is: If the PTO examiner is going to make such a thorough search, why did you have to make one? The answer is that, in essence, proof of patentability rests with the inventor and not with the PTO. That is a somewhat legalistic interpretation. But from a simpler, more practical point of view, an untold number of inventions never reach the Patent and Trademark Office because upon making the search, the inventor realizes someone preceded him. If inventors were not required to make a search on their own, the PTO would be burdened with thousands, perhaps millions, of applications that never should have gotten to them in the first place. More importantly, an application for a patent rests on the inventor's sworn statement that he believes his invention is unique. How could anyone make such a sworn statement without first determining, to the best of his or her ability, that the invention is, in fact, unique?

Now begins the process of *prosecution*. This term refers to the processing of your application through the patent office. The examiner may reject one or two of your claims, either by altering or by eliminating them. The prosecution of a patent may take two years or less or it may be a long, drawn-out affair. But on the average, a patent is issued between two and three years from the time the application is filed.

Foreign Patents

Your patent gives you exclusive rights to your invention only in the United States. For similar protection in foreign countries, you must apply for patents in those countries. Virtually every nation in the world has a system for patents, and it is almost always necessary to make an application in each country separately.

If you want to patent your invention in foreign countries, you are strongly urged to consult a patent attorney. If the attorney is himself involved in international patents, he will be able to advise you on the patent procedures and regulations in most major countries. If his expertise is limited to American patents, he probably works through an associate in the country in which you want to patent your invention.

Variations in the patent regulations of other countries are many. For example, many governments have exceptions to the patentability of certain kinds of items, such as food and drug products. Some countries will grant patents only for a product; others will grant them only for a manufacturing process; still others may grant patents for both process and product.

The definition of novelty also varies. Most countries do not regard an invention as novel if the product has been used or written about before filing the patent application. But use and publication may mean only within the particular country or they may mean within the entire world. Reciprocal agreements make exceptions for inventions patented—or with patents pending—in friendly nations.

In general, American inventors can expect to be treated on the same level as nationals. That means you, as a foreigner, will not be subjected to any rules and regulations that do not also apply to local citizens. The United States has agreements with approximately 100 countries that make such treatment possible. As a result of some of these agreements, some countries, which require an inventor to work for a specified period in that country, have waived the requirement for Americans.

The important thing to remember is that the various bilateral and multilateral agreements in which the United States participates help make it a little easier for an American to obtain the patent in a foreign country. *It does not mean that your American patent protects you overseas.*

Not surprisingly, many of the complications concerning foreign patents are in the process of being ironed out by an organization known as the World Intellectual Property Organization (WIPO), or, by its French name, *Organisation Mondiale de la Propriété Intellectuelle.* WIPO is an international organization charged with implementing and enforcing the Patent Cooperation Treaty (PCT). At this writing, the details of how WIPO works are yet to be fixed.

All that is known now is that by 1980, it should be possible to file an international patent application with the United States Patent and Trademark Office in Washington. "The effect of the international application," says WIPO, "is the same as if national applications had been concurrently filed with the national patent offices . . . of those countries party to the PCT which the applicant designates. The international application is then subjected to a search of the 'prior art' by the United States Patent and Trademark Office and the applicant is placed in a position in which he can decide, on the basis of the international search report, whether it is worthwhile to pursue his application in the various countries he has designated." As of June 1, 1978, the countries party to the Patent Cooperation Treaty are the United States, the Federal Republic of Germany, Switzerland, "and probably several other highly industrialized countries such as France and the Soviet Union." About twenty countries were expected to be party to the PCT by June 1, 1978. Japan was expected to join later that year.

WIPO may prove to be the answer to many of the problems surrounding foreign patents. They will be glad to send you all available information if you write to them at 32, Chemin des Colombettes, 1211 Geneva, Switzerland. WIPO has promised to have detailed brochures available free of charge. Place your orders now.

Ultimately, if you do have a patentable invention and all hurdles have been conquered, you will receive from the United States Government a gorgeously rendered document called Letters Patent, declaring to the world that for seventeen years you have exclusive rights to the invention described in that document. You have your patent.

Now, what are you going to do with it?

CHAPTER

5

Great Expectations

If a man write a better book, preach a better sermon, or make a better mousetrap than his neighbor, though he build his house in the woods, the world will make a beaten path to his door.

Of all the pithy observations of the witty and erudite Ralph Waldo Emerson, the one quoted above is probably the most famous. It is also probably the silliest. It should be obvious, even to the most innocent and faithful, that before the world can do any path beating, it must know about the existence of that better mousetrap. In its own conservative way, the PTO recognizes that fact: in addition to listing the patents of the week in the *Official Gazette*, the PTO also lists patents that are available for assignment or licensing.

Before we discuss the various ways in which you can turn your patent into financial rewards, let's dwell for a few minutes on the kinds of rewards you can expect. As stated earlier, the myth of ordinary citizens becoming overnight millionaires because of an invention is exactly that—a myth. Issuance of a patent is not a signal to quit your job, put a down payment on a mansion, make serious visits to the nearest Rolls Royce showroom, or plan a world cruise. With a little luck and a lot of perseverance, you should be able to supplement your present income enough to acquire some of the things you may have had to do without until now. With a great deal of hard work and conscientiousness, you may even derive a good livelihood from your invention.

For some inventors, the satisfaction of knowing they have devised a patentable item is enough and they are content to have the beautiful patent document adorn a wall in their home. Most inventors, however, find even greater satisfaction in watching their bank accounts grow, and there are several ways to accomplish this.

With few exceptions, the greatest financial reward can probably be achieved by manufacturing and marketing your invention yourself. But this also involves the greatest amount of work and the most risk. Going into business for yourself, with your invention as the foundation of that business, is not easy. We have, therefore, reserved a detailed discussion of how to do this for the next chapter.

A patent is personal property, which may be sold or mortgaged to someone else. It may even be bequeathed to someone else in a will. Like any other piece of personal property, a patent may be jointly owned and a portion of it—say, a half or a quarter interest—may be sold.

You may also mortgage your patent. In that case, the "patent property" is owned by the mortgagee or lender until the mortgage or loan is repaid, at which time the patent is transferred back to its original owner. "An assignment, grant, or conveyance of any patent or application of patent," advises the PTO, "should be acknowledged before a notary public or office authorized to administer oaths or perform notarial acts. The certificate of such acknowledgement constitutes *prima facie* evidence of the execution of the assignment, grant or conveyance." Once the instrument is executed, it should be sent to the Patent and Trademark Office where it will be included into the records of the patent. If you do not yet have someone who is interested in buying or licensing your patent, notify the Patent and Trademark Office that your patent is available for such purposes and the PTO will include that information in the *Official Gazette*.

Selling Your Patent

Think carefully before selling your patent outright. The purchaser of your patent will undoubtedly limit his purchase price to the amount of money he feels he can afford to lose. For most speculators, whether individuals with a gambling instinct or giant corporations with considerable financial resources, that estimate tends to be conservative, to say the least. The money you get from the sale is all you will ever realize from your investment.

Some factors that might influence an outright sale are:

1. *Desperation.* An outright sale of your patent may be the only recourse left after all other alternatives for making money from it have been exhausted.
2. *Expediency.* If your career as an inventor is being impeded by some problem that can be easily solved by a quick dose of ready cash, it may be worth your while to give up your invention now, get the immediate problem out of the way, and go on to bigger and better things.
3. *Timeliness.* Your invention may have a built-in time limitation. Perhaps it involves some aspect of current events. It may be a temporary stopping point along the way to rapidly developing technology. It may relate to some hot and lively fad, which shows signs of eventually fading. If, for example, you have developed a device for use in conjunction with CB radios, you might be

well-advised to unload the patent as quickly as possible onto a company capable of manufacturing the item and getting it onto the market in a hurry, because all indications show that the CB radio craze has, at best, leveled off.

Royalties

Another way of selling or assigning your patent is to do so on a royalty basis. In all likelihood, if there is some market interest in your invention, a royalty arrangement will be much more interesting and acceptable to potential manufacturers and purveyors of your product. In general, the company taking the assignment of your patent on such a basis will provide a small but respectable sum of money to seal the bargain. Typically, that lump sum is an advance against future royalties. Then, once the product is being manufactured and sold, you, as the inventor, will receive either an amount equal to a percentage of gross or net sales, or a fixed amount for each unit sold. These royalties will be applied against the advance paid to you until they equal the advance. You will then continue to receive additional royalties.

Invariably, it will take longer for you to receive your royalties than it would to obtain the cash from an outright sale, but in the long run, you are likely to make more money.

Licensing

A variation of this method is licensing. Here, you retain ownership of the patent, but grant permission to one or several other parties to "practice" it. The exact nature of the licensing arrangement depends on a number of factors.

You may decide to divide the country into territories and

assign exclusive rights to your patent on a regional basis, with a different licensee in each region. A perfect example of this arrangement is the soft drink industry. National brands of soft drinks, such as Coca-Cola and Seven-Up, are produced regionally by independent bottling companies, under license by the parent company.

If your product warrants it, you may want to license the patent in accordance with trade divisions. One company might be given a license to market your invention to industry, while another would handle retail or consumer sales and distribution.

Still another limiting factor in licensing could be time. If you decide a licensee should be able to succeed with your product within a prescribed period, you may put such a time restriction on your license.

There are many fine points involved in licensing or selling your invention on a royalty basis. After you find an interested party, it is a good idea to get your lawyer involved in the proceedings.

Of course, you may choose to do nothing at all with your patent. (Another myth that persists is that deep down in basement laboratories of giant corporations, there are scientists who produce inventions, obtain patents for them, and then bury the patent for the sole purpose of keeping the product off the market. You have no doubt heard the story of the major oil company whose research and development department has come up with a simple tablet—chemical composition unknown—that can convert a tankful of water to gasoline. There are of course nuisance patents, which exist for the sole purpose of thwarting competitors, but they are few and far between.) There is no real harm in allowing the patent to lie dormant. But obviously, the longer a patent remains unexploited, the less it is likely to be worth. Other patents for similar products, which are improvements on or departures from your invention, are almost certain to come along and virtually cancel out your dormant patent. Thus, while you have the protection of your patent, you have little else, because others probably will have circumvented your protected invention with something better.

Perhaps the only good that can come from an unused patent is that after seventeen years, when the patent expires, it becomes a tax deduction. Circumstances vary with individual cases and a knowledgeable accountant should be consulted. In general, however, tax laws allow depreciation on the total cost of the patent, beginning with the date the patent was granted. Costs include PTO fees, the registered practitioner's fees, and the costs of drawings, experiments, tests, etc., that the inventor paid to develop his invention. Incidentally, the complexities of the tax laws are yet another good reason for keeping complete, careful and detailed records. Should you decide to deduct the cost of your invention, you may be called upon to produce substantiating evidence.

But surely you did not become an inventor for the sole purpose of having a tax deduction available. One of the reasons for producing an invention is to make money from it. As we said earlier, you can take your invention and go into business with it, or you can offer it to someone else to make a profit and share that profit with you. The next two chapters examine these alternatives in detail.

CHAPTER

6

Going It Alone

The home permanent, the ball-point pen, and the Xerox copier are three well-known examples of hundreds of inventions brought to market by individuals with enough courage, foresight, and ingenuity to set up their own businesses. They either built huge, highly profitable corporations out of those small businesses or ultimately sold them to large companies for vast sums of money.

If Edwin Herbert Land could do it with his Polaroid camera, you can do it too—maybe. The two most important factors are, of course, you and your invention. To begin let's take a look at your invention.

Appraising the Invention

The first step you must take is often the most difficult. Your invention is a creation of your imagination and, as such, it is a little like your baby. Just as every parent is convinced that his child is cute, intelligent, and an altogether wonderful human being, virtually every inventor believes his invention is unique and useful, and desired: there are millions of people out there just waiting to get their hands on a product exactly like the one he's invented. But belief is an emotional condition. It is time to be cold, hard, and objective. You must ask yourself, on the basis of your knowledge, experience, and, yes, even your intuition, whether there really is a market for your invention.

Are there really enough people in the world of commerce who are seven feet tall, who are amateur photographers, and who either have an existing need for our tripod or who can be convinced they ought to have it? Can we be either absolutely or even reasonably certain that other tripods already on the market—tripods that are perhaps cheaper, certainly better known, and already enjoying the benefits of a brand name, an advertising program, and a distribution setup—do not already satisfy whatever market exists?

At the very beginning of this book, we talked about a shopping cart designed to climb stairs. That shopping cart is covered by Patent No. 3,420,540, issued to William Bird of New York City on January 7, 1969. After 8 years, Mr. Bird was still looking for someone to pick up his invention. In his promotional literature, the inventor claimed that this cart would be "a good seller in a world market, especially in large cities where there are large urban areas comprised of two- and three-family walk-up apartment houses . . . even in elevator apartments there are entrance and lobby stairs to climb. The cart would particularly appeal to the elderly and the partially infirm to help with their shopping and laundry needs." Anyone who has ever had to handle a shopping cart can certainly see the wisdom in Mr. Bird's suppositions. Apparently, however, at this writing, the shopping cart market has not agreed.

Tests and Evaluations

The next serious question to be answered is: Will the product withstand the rigors of daily use? No matter how thoroughly you, your family and your friends use and abuse the prototype of your invention, chances are you will never be able to duplicate actual use conditions. A testing laboratory, on the other hand, can. Commercial testing laboratories can be found in most large cities by consulting the Yellow Pages. If you use a patent agent or patent attoney, he can probably put you in contact with a reliable laboratory. Once the lab gets your product in its clutches, tests will be made on construction, material, and durability. The testing laboratory will also compare your product with those of your competitors and give you a detailed report on what makes your product better or worse. Frequently, a testing laboratory will not only tell you what is right and what is wrong with the product, but will also help you correct it by making recommendations.

Check with your local librarian or the nearest field office of the U.S. Department of Commerce for directories of testing and research laboratories. Of course, independent testing laboratories charge for their services. But in the long run it is usually worth the cost since their findings can save you much more burdensome expenses later.

Ideally, it would be useful to have your invention evaluated before you go through the trouble and expense of applying for a patent. But as we have seen, that presents the danger of the invention becoming known, which automatically removes it from patentability. There are, however, two ways of getting an evaluation before applying for your patent. If your invention is related to energy, the National Bureau of Standards will give you a free evaluation. You can obtain an Evaluation Request Form (NBS-1019) and general information by writing to the Office of Energy-Related

Inventions, National Bureau of Standards, Washington, D.C. 20234. No matter what your invention covers, you can obtain a detailed evaluation for $25 from the Experimental Center for the Advancement of Invention and Innovation, College of Business Administration, University of Oregon, Eugene, Oregon 97401. Write for information before sending them your invention.

Market Research

If you are convinced your invention is marketable and you know it will work as it is supposed to, your next step is to conduct some market research. Once you have your patent, you may talk about your invention freely. By all means, talk about it to friends and acquaintances who might have some real use or interest in your product. If you don't already know, you will soon find that when you ask for an opinion, you are going to get one, regardless of whether the person giving it has any expertise. Confine your conversations, therefore, to people who are likely to know. There is no point in discussing an automotive accessory with someone who doesn't even have a driver's license, just as there is no point in discussing a baby-care article with a confirmed bachelor.

Talk to retailers likely to sell your product or, if applicable, industrial establishments likely to buy your invention. Such conversations can give some indication, however unstructured and random, of how the market feels about your product.

Talk to people who manufacture or sell a related but noncompetitive product. If, for example, you have invented a new kind of shoe tree, you might get some excellent insights into the market through discussions with shoe manufacturers, importers and wholesalers.

Once you have a good feeling about the existence of a market for your invention, you should invest in a real market survey. The word "invest" is used deliberately: market surveys cost money. How much money depends on (1) the money you have, (2) the money you are willing to spend (3) the nature of the invention, and (4) the nature of the potential market. For some products, such as an easy-to-use needle threader, a market study can probably be made for as little as $100 or $150. For a device that provides greater stability to aircraft carriers, a market study could run into the millions. Nevertheless, if you have decided to promote your invention on your own, you must set aside a portion of your capital for some kind of a market study. Bear in mind, however, that whatever your market study ends up costing, it can never equal the cost in dollars and simple aggravation of manufacturing an item in large quantity, only to discover that nobody wants to buy it. One of the saddest stories ever told is of an inventor who designed a beautifully simple plastic holder for trading stamps and trading-stamp albums when almost every retailer was giving away such stamps. Unfortunately, he somehow did not remember to take the measurements of a typical trading-stamp book and manufactured thousands of plastic holders too small to hold the items for which they were intended. One cannot help but wonder how it was possible to make such a mistake. Since the oversight was correctable, had the inventor conducted a simple market study, he would have seen his mistake immediately and been able to change it. As things turned out, he put all his funds into manufacturing before bothering to test the market and was unable to do anything but stare at his pile of useless trading-stamp holders and ponder about what might have been.

How long should your market study run? That depends to a great extent on the product. A market study for an item with rapid turnover—a new kind of household glue, one-time carbon paper for office use, a dishwashing product—requires much shorter tests than something more durable, such as a typewriter, a can opener, or a tripod designed for seven-foot-tall amateur photographers.

Where should you conduct your market tests? If at all possible, select an area close to home so the test can be supervised with a minimum of travel or telephone expense and without the need to hire the services of other people. Bear in mind, however, that proximity can sometimes be a hindrance. If you live in Southern California and have invented a new type of earmuff, you will have to travel some distance to conduct the market survey. If you live in Arizona and have invented a special kind of umbrella, you will have a difficult time test-marketing the product in Phoenix.

Try to select a city or region large enough to give a good cross-section of purchasers, but one that is not too large. Contrary to popular belief, New York City is one of the worst places to test-market a product.

Try to select a test market that more or less approximates the national market you are going after. If your product has an ethnic appeal, select a test market containing a high percentage of the ethnic group you are trying to reach. If the product has a regional appeal (climate, proximity to water, etc.) choose an area similar to the regions where your national sales would be. In other words, if you have something that will best be sold during the winter months, select a test market that experiences an actual winter season.

Despite Ralph Waldo Emerson's declamation about better mousetraps and path-beating, you must advertise in your test-market area. Therefore, choose a region with enough advertising media to get your product before the public eye, but not so much that the cost will be more than you can handle. Most advertising rates are based on "cost-per-thousand." The cost of a quantity of space in a newspaper or periodical, or the cost of a quantity of time on a radio or television station, is broken down to what it costs you for each thousand people you reach. That can be a deceptive figure and must be watched carefully. The advertising rate for a newspaper in a city like New York, Chicago, or Los Angeles is likely to be cheaper on a cost-per-thousand basis than it would be for a city like Duluth, Des Moines, or

Mobile. But you may be reaching more people than you want to in the larger cities and the overall *total* cost of advertising in those large cities will be more than the same space or time would cost in a smaller city.

While we're on the subject, unless you have a tremendous amount of money to spend on advertising, concentrate on the so-called "print" media (newspapers and magazines) rather than the "air" media (radio and television). A message on a printed page has a much longer-lasting impact than a message over radio or television. To be effective, radio and television advertising require constant—and therefore expensive—repetition. The only exceptions, generally, are products where sound is an important selling feature (such as phonograph records) or when a prospective customer must see a product in action to fully appreciate its desirability. In the latter case, television can be a highly successful advertising medium. In most cases, however, stick to print.

If your advertising requires some direct action or response from the consumer—other than simply going into a store and buying the product—you should code the ads so you can tell which ones generate the most business. For example, if you ask the consumer to write for a brochure, to order the merchandise through the mail, to take a special discount coupon into the store, or to bring in the ad and exchange it for a premium, you will have the ad, or a portion of it, back in your hands. By coding the portion to be returned, you will identify the source of the ad and see which ones "pulled" best. Sometimes the code will consist of a letter or number printed unobtrusively in a corner of the ad. Most of the time, however, the code will be a "department" and will not be difficult to decipher. "Department NYT-4-9" might be used for a coupon published in the New York *Times* on April 9th. The same coupon, coded "Department CPD-617," might be from the Cleveland *Plain Dealer* of June 17th. It will not be necessary to get fancy or tricky with codes. A simple series of numbers will do, provided you keep a list of those numbers and the individual advertising media to which they refer.

(These basic rules apply not only to the test-market phase of your business, but to your full-fledged, fully-budgeted advertising program.)

You must, of course, keep careful records of your market test. Some of the things you will want to keep a very close watch on are:

1. *Volume of sales.* How many units were sold during the test period?
2. *Number of units per period of time.* How many were sold in a day; in a week? Does the product sell better on weekdays or weekends? Does it sell better at certain times of the day than at others?
3. *Customers.* Who is actually buying the product? Try to get a detailed breakdown by age, income level, occupation, neighborhoods, etc.
4. *Outlets.* Is the product selling better in centrally located department stores, in stores at shopping centers, or in neighborhood hardware, grocery, or specialty shops?
5. *Customer awareness.* Was the purchase motivated by an advertisement, by a sales person, or by recommendation of a previous purchaser?

The easiest and most efficient method of obtaining the answers to questions 3, 4, and 5 is to enclose a simple, straightforward questionnaire with your product. It should be short enough to fit on a postcard (folded, if necessary) inserted into the package. If you have ever filled out a card to "register" a product for coverage by the manufacturer's warranty, you know how often companies obtain market data this way.

How Much Should You Charge?

The price you set for your product has three important effects, two of which are obvious. *First,* if your price is too high, you will discourage buyers. *Second,* if your price is too low, you will fail to realize a profit. And *third*—much more subtle but just as important as the first two—the price you set for a particular product will often say something to the prospective buyer about the quality of that product.

"Quality" is a difficult word to define. Of course it includes such characteristics as workmanship, durability, performance and appearance. But quality also suggests some intangible characteristics to certain buyers for certain products: prestige, distinction, an air of luxury or affluence. After all, with disposable butane cigarette lighters selling for $1.49 and having virtually no mechanical flaws, why is there a market for 14-carat gold cigarette lighters selling for ten, twenty, and thirty times as much? No doubt you have, on more than one occasion, rejected purchasing some item because the price was too low, on the premise that "if it's that cheap it can't be much good."

Similarly, a product can be overpriced even if it is in the affordable range of most consumers. An ordinary wooden pencil would be difficult to sell for fifty or sixty cents. There are some practical aspects to pricing, which we'll get into in a moment, but remember that when you establish a price for your product, the price can often be as much of an aspect of your product's image as the package and the product itself. Again, take a look at what the competition is doing. If a tripod similar to our hypothetical invention sells for $35, we might, on the basis of the special applications of our tripod, charge perhaps $40 or $45. But if we were to charge $75 or $80, even if the product were worth it and our potential customers could afford it (amateur photographers pay a great deal more than that for camera bodies, lenses, and other accessories), the average prospect may say to himself: "That's too much money for a tripod." He will probably not say: "That's more money than I have to spend." In this case, what he can afford becomes immaterial. What is relevant is how the price compares with similar products already on the market.

The first basic rule in setting a price for your product is to recover your cost of producing that product. Sometimes there are legitimate arguments for temporarily sustaining a loss in order to get a new item into the market place, but that is a game to be played only by businesses with huge financial resources. You must carefully determine your cost per unit. Everything involved in getting that item into the hands of the consumer must be calculated: the cost of materials, production, and packaging; the cost of developing and patenting the item; the cost of transportation and distribution; the licensing fees, taxes, and professional fees (e.g., bookkeeping, accounting, legal); the overhead—rent, electricity, telephone, stationery, postage—all ongoing, fixed costs required to set up your business and keep it operating; and the cost of labor.

Labor costs are always a sensitive issue. Certainly you must figure in the expenses of hiring the services of other people to work for you whether on a full-time, part-time, or contractual basis. But many people going into business prefer not to include the cost of their own time or the time their friends and relatives are willing to donate to the enterprise. They think that in many instances they wouldn't be paid for the time anyway. If you run your business in your spare time, you may reason that you lose nothing by excluding the cost of that time as part of your expenses. After all, no one pays you for pursuing a hobby, watching television, walking the dog, or taking a nap after dinner. From a practical point of view, however, not calculating the cost of "free" labor is, in the long run, a false economy. If your business succeeds, as you expect it will, eventually that labor will cease to be free. When that happens, you will be faced with the difficult and uncomfortable task of having to rework your pricing policy. Either you will have to increase the price of your product or you will have to reduce other costs to make up for the no-longer-free labor. So even if you don't draw income from your business at first, it is always a good idea to calculate your costs as if you were. The money allocated for temporarily-free labor can and should be reinvested in the business to enhance stability and accelerate expansion until the labor stops being free.

The complexities of calculating cost-per-unit, as well as other financial aspects of setting up a business, can be greatly simplified by getting help from the Small Business Administration. Even if you are the type who prefers to avoid consulting experts (a risky attitude for *any* businessman), you should at least avail yourself of the SBA's literature. For example, you can obtain, free of charge, such SBA publications as: *What Is the Best Selling Price?; Business Plan for Small Manufacturers; Basic Budgets for Profit Planning; Pricing for Small Manufacturers; Are You Kidding Yourself About Your Profits?; A Pricing Checklist for Managers,* and *Keeping Records in a Small Business.* At nominal cost, the SBA offers *Cost Accounting for Small Manufacturers* ($1.60); *Decision Points in Developing New Products* (90¢); *Managing for Profits* ($1.90), and scores of other useful pamphlets, folders, and books. (Note: prices are subject to change, but not by very much.)

Break-Even

It is absolutely essential that you determine, before attempting to sell a single piece of merchandise, what your "break-even" point will be. *The break-even point is the point at which you have done enough business to exactly equal the amount of money you have spent to do that much busi-*

ness. There are a variety of formulas for calculating the break-even point and it is difficult to specify in this book which method is best without pertinent details about the particular business. It is therefore strongly recommended that you discuss working out your break-even point with your accountant or banker, or any of the other agencies mentioned at the end of this chapter. These agencies are glad to help you, often without charge. But even if you must pay someone to formulate your break-even point, get it done. (Just remember to include the cost of that service when calculating your overall costs.)

On pages 62 and 63 are sample break-even charts developed by the Small Business Administration. They are shown here only as examples; whether they are right for your particular business depends, as mentioned above, on the particulars of that business. These illustrations are from the SBA's booklet, *Guides for Profit Planning* (85¢).

Now that you have a good idea of what your minimum charges will be, you can consider pricing as a marketing or sales strategy. In general, there are two types of pricing policies: *skimming* and *penetration.* (There are, of course, circumstances in which variations and combinations of skimming and penetration are useful, but for purposes of this discussion, we will consider the two policies as standing alone. Your own particular product and marketing circumstances will indicate what adjustments and variations should be made.)

Skimming involves, in effect, skimming the cream off the market. Skimming works best with a product so new and unusual that people will want to buy it regardless of the price (within reason, of course). When you have such a product, for which there is no competition, hard-hitting promotion and effective salesmanship count for more than the item's cost. By zeroing in on the people who will want your product and who will be able to afford it, you may be narrowing your market potential somewhat. It doesn't matter;

CONVENTIONAL BREAK-EVEN CHART

Expense and Income
(in thousands of dollars)

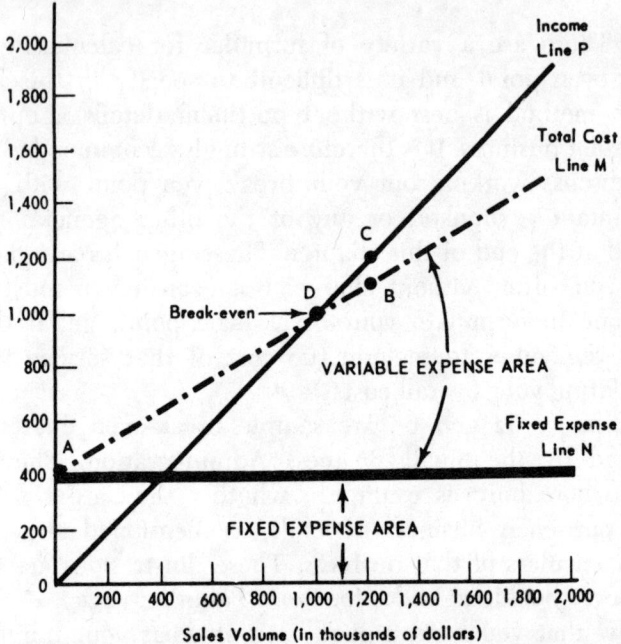

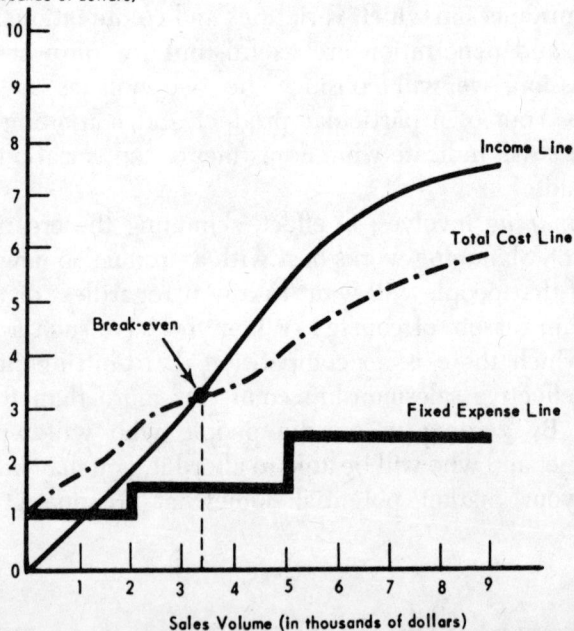

Income
Line P

Total Cost
Line M

2,000
1,800
1,600
1,400
1,200
1,000
800
600
400
200
0

C

D

B

Break-even

VARIABLE EXPENSE AREA

Fixed Expense
Line N

A

FIXED EXPENSE AREA

200 400 600 800 1,000 1,200 1,400 1,600 1,800 2,000

Sales Volume (in thousands of dollars)

NONCONVENTIONAL BREAK-EVEN CHART

Expense and Income
(In thousands of dollars)

10
9
8
7
6
5
4
3
2
1
0

Income Line

Total Cost Line

Break-even

Fixed Expense Line

1 2 3 4 5 6 7 8 9

Sales Volume (in thousands of dollars)

BREAK-EVEN CHARTS WITH ALTERNATE BASE LINES

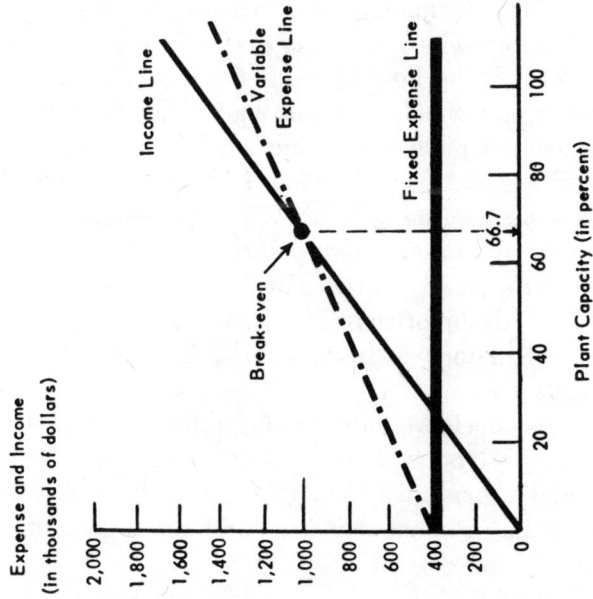

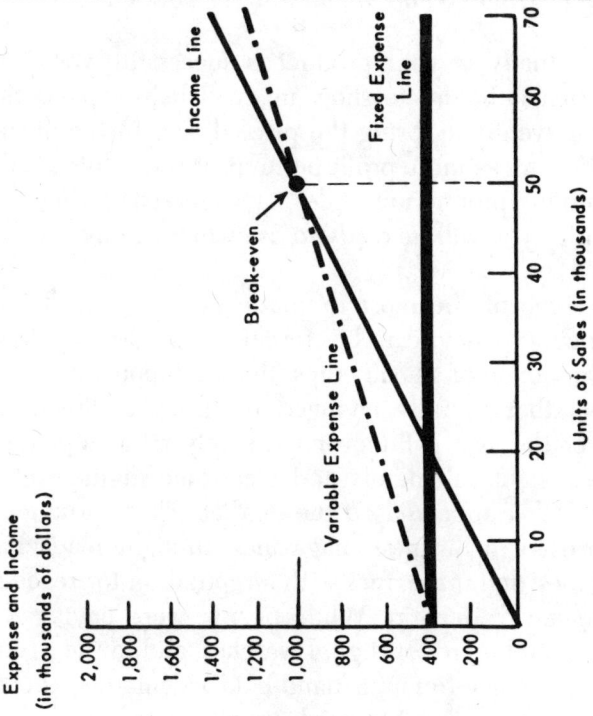

eventually, if your product is successful, your competition, which is bound to show up, will help expand that market and eventually bring the price down. In the meantime, you can achieve more profit per unit of sale while simultaneously gaining production, sales, and merchandising experience, which you will be ready to use when you move into the mass market.

Some of the most dramatic examples of skimming as a pricing policy can be found in products related to the production of silicon chips, those fabulous little semiconductors that literally changed electronic technology. Anyone over the age of fifteen must surely remember the prices of hand-held calculators and electronic digital watches when they first appeared on the market. These products were advertised in business magazines, in-flight magazines on airplanes, and magazines with a reputation for reaching a high-income readership. While people were paying $100, $200, $300 and more for digital watches, and in the $100 range for four- or five-function hand-held calculators, the companies manufacturing these products were hard at work improving production technology and bringing down the cost of semiconductors. The result was that an excellent hand-held calculator can be bought today for less than a good necktie, and digital watches can be had more cheaply than dinner for two in a moderate restaurant.

Even as the prices of hand-held calculators were tumbling, ads were beginning to show up in in-flight magazines and elsewhere for a ballpoint pen that incorporates an electronic calculator in its barrel. This item was promoted as being well worth the price of $89.95. Within a couple of years, the pen calculator was being offered in mail order catalogs for $19.95.

If you decide on a skimming price policy, remember that eventually you will be moving into the mass market. It is usually a good idea to maintain some kind of difference between the product that is sold at the higher price and the one sold at the mass-market price. That difference may be one of design or of nonessential material. Sometimes a simple substitution of metal for plastic will do the trick. Or a little gold plating may give the higher-priced item a somewhat richer look. In the case of our tripod, if we want to go with a skimming policy, we might decide to make the higher-priced model a "nonreflecting black" color, or supply it with a "deluxe" carrying case.

If the price you set proves to be too high, you can always reduce it, possibly even in stages. That's another advantage of skimming: it's much easier to begin with a high price and lower it than it is to increase a price set too low initially.

But skimming is not always possible. Sometimes conditions exist which eliminate any possibility of a skimming policy. If the unit price of a product is, by the very nature of the product, so high that there cannot be stratification of the potential market, then skimming is ruled out. At the opposite extreme, a "skimmable" market may not exist because a product is so commonplace that everyone uses it. Generally, in such circumstances, there is enough competition around to effectively preclude a skimming policy. At such times, we should consider the other side of the pricing coin: penetration.

Penetration is a pricing policy designed to get your product into the mass market fast and deep. If it looks as though the competition might hound your heels shortly after your product appears, keeping your prices as low as possible will help discourage the competitors. It may not be worth the time, effort, and cost for them to chase after you.

The cost of production may also influence your decision about whether to go for penetration pricing. Often, a larger volume of production reduces the per-unit production costs. Certainly, it reduces the overhead costs per unit. Your rent remains constant whether you produce a hundred units per month or a thousand. But the amount of rent calculated per unit is obviously lower when, within a fixed period, more units are produced. That same principle applies to distribution. If the item you are manufacturing has a shipping weight of 1 pound, and the transportation company has a minimum rate for 100 pounds, your transportation will cost the same whether you ship 1, 50, or 100 units.

Whether you decide on a skimming policy or a penetration policy to calculate your prices, or something in between, remember that different pricing policies usually require different promotion and sales techniques and may also require different distribution methods. If you decide on a skimming policy, there seems little point to advertising in media that reach a high percentage of low- or middle-income families. Nor would you have much success in trying to sell your product through discount stores or bargain outlets. If you decide on a penetration pricing policy, an advertisement in *The New Yorker* or *Gentleman's Quarterly* will not go very far toward reaching your market. If you confine your distribution to exclusive department stores or specialty boutiques, your sales will be far short of your goal.

To summarize, here are the points you should carefully consider in determining a price for your product:

1. How much will it cost you to manufacture and sell your product?
2. What is the likeliest market for your product? Consider both the immediate potential market and the long-range market.
3. What is the nature of your competition? Take a hard look at what the existing competition is and analyze what future competition is likely to be.
4. What are the best ways to distribute your product? Consider all alternative possibilities to determine whether you can effectively and legitimately establish several price levels. Perhaps you can distribute directly to the ultimate consumer through mail order. You may have industrial or commercial buyers as well as con-

sumers. With some modifications, your product may do as well in high-priced retail outlets as in mass-market outlets.

Above all, whatever pricing policy you decide on, always make sure your price covers your costs. That may seem like an obvious rule to follow, but the number of businesses that have failed by neglecting to observe that simple injunction is appalling.

Going Into Business

There is nothing to compare with being your own boss. Running your own business gives you independence, a sense of pride, the right to make all major decisions, the opportunity to try new ideas without fighting your way through tangles of red tape, and perhaps most important, the kind of contentment that comes with the knowledge that you are working for yourself and your own betterment rather than for the benefit of someone else.

But for each of these advantages, there are also disadvantages. Especially in a new, small business, no one works harder, keeps longer hours, or takes greater risks than the boss. If you have the right to make major decisions, you will also be the only one to take responsibility for any decisions that turn out to be wrong. If you stand to benefit most from the profits of that business, you will also run the greatest risk of financial loss. Along with independence comes responsibility—meeting a payroll, paying the rent, paying the taxes, and satisfying the creditors. Even if you do everything right, circumstances over which you have no control, such as a sudden change in the economy or a massive attack by competitors, will have to be faced, and sometimes suffered, by you and you alone.

Even the idea that you are your own boss is not entirely

accurate. For one thing, your customers will dictate or influence many of your actions and decisions. Government agencies and insurance companies concerned with health, safety, and other regulations will also influence or dictate to you.

Nevertheless, if your business is based on a quality product, if you offer genuine value, and if you exercise sound management practices, you stand a good chance of succeeding.

Are You the Business Type?

Before you even decide whether to go into business for yourself, you must make one very important decision—a decision with which other people who know you well can help, but which ultimately rests solely with you. The question that you and you alone must answer is: Are you the type to go into business for yourself? Be objective. Try to imagine yourself as two separate people, the person doing the hiring and the person applying for the job of managing a business. Would you hire yourself? After all, no one will be more important to your business than you. That's why you should interview yourself carefully, covering all the points you feel are necessary for running your business. If you fail on a point or two, that does not necessarily spell disaster. You may not be top-notch at dealing with details and keeping records. It doesn't matter, so long as you recognize the failing and compensate for it either by exerting an extra amount of effort in the area in which there is a lack, or by hiring someone who can make up for your deficiency.

On pages 69-71, is a checklist based on a worksheet prepared by the Small Business Administration. Look it over and put a check mark next to the answers you think most closely reflect the truth about yourself. Be brutally honest in completing this questionnaire. Don't rush to check off an answer to a question you are unsure of. Give it some thought and, if necessary, discuss it with your spouse or a close friend or relative. Remember that no one is going to grade you on this checklist and no one will see it but you. It might be one of the most important steps you take in deciding whether to go into business for yourself.

Are You the Business Type?

This checklist is adapted from the Small Business Administration's publication, *Checklist for Going Into Business*.

Under each question, check the answer that says what you feel or comes closest to it. Be honest with yourself.

Are you a self-starter?
☐ I do things on my own. Nobody has to tell me to get going.
☐ If someone gets me started, I keep going all right.
☐ Easy does it, man. I don't put myself out until I have to.

How do you feel about other people?
☐ I like people. I can get along with just about anybody.
☐ I have plenty of friends—I don't need anyone else.
☐ Most people bug me.

Can you lead others?
☐ I can get most people to go along when I start something.
☐ I can give the orders if someone tells me what we should do.
☐ I let someone else get things moving. Then I go along if I like it.

Can you take responsibility?
☐ I like to take charge of things and see them through.
☐ I'll take over if I have to, but I'd rather let someone else be responsible.
☐ There's always some eager beaver around wanting to show how smart he is. I say let him.

How good an organizer are you?

☐ I like to have a plan before I start. I'm usually the one to get things lined up when the gang wants to do something.

☐ I do all right unless things get too goofed up. Then I cop out.

☐ You get all set and then something comes along and blows the whole bag. So I just take things as they come.

How good a worker are you?

☐ I can keep going as long as I need to. I don't mind working hard for something I want.

☐ I'll work hard for a while, but when I've had enough, that's it, man!

☐ I can't see that hard work gets you anywhere.

Can you make decisions?

☐ I can make up my mind in a hurry if I have to. It usually turns out O.K., too.

☐ I can if I have plenty of time. If I have to make up my mind fast, I think later I should have decided the other way.

☐ I don't like to be the one who has to decide things. I'd probably blow it.

Can people trust what you say?

☐ You bet they can. I don't say things I don't mean.

☐ I try to be on the level most of the time, but sometimes I just say what's easiest.

☐ What's the sweat if the other fellow doesn't know the difference?

Can you stick with it?

☐ If I make up my mind to do something, I don't let *anything* stop me.

☐ I usually finish what I start—if it doesn't get fouled up.

☐ If it doesn't go right away, I turn off. Why beat your brains out?

How good is your health?

☐ Man, I *never* run down!

☐ I have enough energy for most things I want to do.

☐ I run out of juice sooner than most of my friends seem to.

Now count the checks you made.

How many checks are there beside the *first* answer to
each question? ____

How many checks are there beside the *second* answer
to each question? ____

How many checks are there beside the *third* answer to
each question? ____

If most of your checks are beside the first answers, you probably have what it takes to run a business. If not, you're likely to
have more trouble than you can handle by yourself. Better find a
partner who is strong on the points you're weak on. If many
checks are beside the third answer, not even a good partner will
be able to shore you up.

After you have completed the checklist, there are several
other factors you should consider about the business you are
planning to enter:

1. *How much do you know about the field?* If you plan to
   manufacture and sell tripods for seven-foot-tall photographers, you ought to know something about cameras,
   something about tripods, something about the characteristics of amateur photographers, and something
   about the retail and mail order markets for photographic
   accessories.

   Ideally, your own work experience is the best training
   you have. But if you have no know-how, then at the
   very least, be sure to acquire some education. How
   much and what kind depends largely on the field you
   are entering. It may be necessary for you to go back to
   school. On the other hand, you may be able to learn a
   great deal by doing a little studying on your own. The
   agencies and organizations discussed toward the end of
   this chapter can go a long way in helping you with
   your self-education.

You should not only learn something about the specific field, but also study some of the principles, techniques, and problems of management. By keeping your eyes and ears open and being constantly on the alert, you can acquire vast amounts of useful information from your job, your colleagues, your supervisors and superiors, the friends you talk to, and the people with whom you do business. Try experimenting with this technique (the success of which never fails to amaze me): see how much management information a total stranger will give you when you simply ask for it. Visit a stationery store, a small restaurant, a community newspaper, or any other kind of business that interests you. Tell whomever is in charge that you are planning to enter a related but noncompetitive business and would welcome any suggestions, advice, or tips he or she can offer. You may get turned down a couple of times, but the chances are better than even that the person you ask will tell you more than you want to know about his or her business. Try it; it really works.

2. *How much expertise is available to you?* One characteristic of a good manager is the ability to surround himself with experts. Get to know the officers in your bank. It is not necessary to have a specific purpose in mind when calling on one of the platform officers in most banks. No doubt the branch manager or one of his assistants would appreciate your choosing a time when the bank is not busy, but they will also appreciate your coming up, introducing yourself, and informing them that you are thinking about going into business and are interested in knowing what services their bank might be able to offer. You will be treated cordially and welcomed with open arms. (If you aren't, find another bank, fast!) You will also probably find out more about a bank's role in small business than you really want to know.

If you don't have a good accountant, start now to look for one. If you know one personally, question him about the services he might be able to provide you as a businessman. If you don't know an accountant, perhaps friends and associates can recommend one or two. If there is a business in your community that you feel is well-run and properly managed, chances are that business is using a good accountant. If the firm is noncompetitive, talk to the office manager, head bookkeeper, or company treasurer, and ask for the name of their accountant.

You should also have a good lawyer. If you don't already have one, you can make contacts with a lawyer in much the same way as you can with an accountant.

If all else fails, check with your librarian for the names, addresses, and telephone numbers of the local associations of certified public accountants and the bar association in your community. Both associations can provide you with names of several practitioners from which you can choose.

3. *Will your business have continuity?* Many enterprising individuals have thrust themselves or their products on the maket, made a tremendous amount of money in a short time, and then gotten out. There is nothing either illegal or immoral about establishing that kind of business, especially if the product is one that suggests a temporary market or that caters to—or even attempts to create—a fad.

But most people go into business with an eye to the future. Before actually setting yourself up in business, think about the long-range possibilities. Is your product likely to become obsolete soon? Will there be a continuing market for it, or would you quickly saturate the market, without much chance of repeat sales? Will there be a continuing need for service, for spare parts, for supplies and accessories? Will the actual structure of your business be adaptable to change, should change be necessary or advisable?

When you have the answers to most of these questions, you are probably ready to go into business. Now, you must decide on the structure of your business.

Types of Business Structure

There are three basic types of businesses you may set up: *single proprietorship, partnership,* and *corporation.* Before we get into the details of these types of businesses, it is very important to remember that any such discussion must deal in generalities. Laws vary from state to state and even from city to city within a state. In some situations, for instance, it is possible to set up a single proprietorship by doing nothing more than hanging out a sign proclaiming to the world that you are in business. In other communities in which all businesses are taxed, it is necessary to file some kind of form before setting up shop. Because of these variations and fine points, it is advisable—and it cannot be repeated too often— to get a lawyer who knows about these various specific regulations. Consider the cost of the lawyer as nothing more than a necessary expense involved in setting up your business, just as rent, advertising and the cost of proper equipment are.

Proprietorship. In general, there are few legal requirements for establishing a single proprietorship. In such a business, you are the sole owner and derive all the benefits from the business's profits. A single proprietorship offers complete flexibility when it comes to management, production, marketing, or other decisions.

The legal costs involved in establishing a single proprietorship usually range from minimal to none. There may be some licensing fees, and you may have to file some forms with local authorities on the collection and payment of taxes. If you choose to do business under some name other than your own, you may have to file a certificate to that effect, usually with the County Clerk, and pay a fee of around $10 or $15.

One of the most important features of a single proprietorship type of business is liability. If you are the sole owner of your business, then any liabilities incurred by that business can be recovered, to the extent of what lawyers call your "personal fortune." That means, if you run heavily into debt and the business cannot repay those debts, you may be required to sell your car, furniture, house, or anything else you own to make good on what you owe. That is why so many business people sign over the ownership of their cars and houses to their spouses.

As stated earlier, all decisions about the business rest with the owner of a single proprietorship. In practice, what that really means is that the *responsibility* for all decisions rests with the owner. If circumstances permit, any intelligent businessman will surround himself with people who can help him make the right decisions. He can also delegate authority to employees. But in a single proprietorship, if a shipping clerk makes a wrong decision that costs the company several hundreds or even thousands of dollars, it is the owner and not the shipping clerk or anyone else who must bear the brunt of that loss.

In terms of longevity, the single proprietorship is the most tenuous: the business dies when the owner does. Of course, provisions can be made in the owner's will to pass the assets of the business to a suitable inheritor. But when that inheritor takes over, it is, for all practical purposes, a new business.

In a single proprietorship, retained net profits are taxed as the owner's personal income.

Partnership. Basically, there are two kinds of partnerships: a *general* partnership and a *limited* partnership. When people talk about partnerships without a qualifying adjective, they usually mean a general partnership.

A general partnership is almost as easy to set up as a single proprietorship.

While you do not need a written agreement to establish a partnership, it is better to have one, not only for records, but also because a written partnership agreement goes a long way in settling any disputes or misunderstandings that may rise when any partner is not a full and equal partner. (In some instances, if two or more people act as partners in a business venture, even though no formal written or oral agreement has been made, under the law they may still be considered partners.)

Most general partnerships assign equal shares of the business to all partners, but it is possible to have senior and junior partners where the junior partners receive a smaller share of the business profits. Junior partners, therefore, are usually expected to provide lesser participation in the business, whether in the form of a smaller financial investment or less time.

A general partnership resembles a single proprietorship in that any net profits the partners derive from the business are taxed as their personal income. Furthermore, as in a single proprietorship, each partner is personally responsible for the liabilities of the business to the extent of his personal fortune. A partnership is automatically dissolved on the death of one partner; however, as in a single proprietorship, a partner may make provisions in his will to pass his share of the business to an inheritor. In many cases, a written partnership agreement specifies the share of the business that is to be passed to a deceased partner's heirs.

A limited partnership is somewhat more difficult to establish. Usually, limited (sometimes called "silent") partners invest in the business by contributing something of value to it: money, premises, materials, etc. Services, knowledge, or expertise are not considered to be legitimate investments for a limited partnership. Limited partners do not participate in the management of the business.

To establish a limited partnership, it is necessary to conform exactly to the laws of the state in which the partnership is organized. This invariably requires filing a contract written in accordance with state laws. That contract limits the liability of all limited partners to the extent of their investments. One requirement of all limited partnerships is that at least one general partner exist. The liability of the general partner is, in limited partnerships, the same as in a single proprietorship or general partnership—to the extent of his personal fortune.

If you decide to establish a limited partnership, be very careful about complying exactly with the state laws. If you fail to do so, then under the law, what you intended to be a limited partnership will be regarded as a general partnership.

The management and administration of a business formed under a general partnership is very much like that of a single proprietorship, except there are two or more heads instead of one. Typically, administrative and management duties are divided among the partners. One partner may be responsible for sales and marketing; another for production and internal administration. (There are still comedians whose repertory of jokes includes garment industry stories about the "inside" partner and the "outside" partner.) Limited partners are not permitted to participate in the management of the business. If they do, they risk being legally considered general partners. However, limited partners can and should be kept fully apprised of all matters concerning the business. They have the right, if not the duty, to examine the books and records from time to time.

Corporation. Under the law, a corporation is a legal entity unto itself. It is, in effect, an "artificial person." This artificial person is created by being given a name and an existence, which are carefully spelled out in the corporation's charter in accordance with the laws of the state in which the corporation is organized. Like a person, a corporation's earnings

—its income—are taxed separately, and its liabilities are limited to the extent of its assets.

The ownership of a corporation is divided into shares of stock. Any number of shares may be issued, in accordance with state laws. Profits not reinvested in the corporation are divided among the shares. If profits for the year are $100 and there are one hundred shares outstanding, each share is allocated $1 in profits. The amount of money each shareholder receives depends on the number of shares he holds. Similarly, in electing corporation officers and directors, each share is entitled to one vote. Therefore, the person holding the most shares controls the corporation. That fact sometimes provides interesting and exciting contests in large, multinational corporations. However, for the small businessman, his family, and his few colleagues, who hold most of the controlling shares, such matters tend to be formalities.

In many instances, there are certain tax advantages to being incorporated. From the standpoint of the small businessman, however, the main advantage in a corporation is limited liability. The corporation may "employ" you as president, general manager, or whatever title "it" decides upon and pay you a salary. But if the company gets into trouble, and creditors come banging at the door, the most one can lose is one's investment in the company. It may become necessary to sell the machinery, hock the typewriters, and auction the furniture the corporation owns, but your own house, car, color TV, and other personal possessions cannot be touched. The only exceptions to that rule are federal taxes and payrolls. When a corporation's assets are not enough to meet taxes and back payrolls, the shareholders may be required to dig into their own pockets to make up the difference.

Generally, management of a corporation rests with a relatively small group of executives charged with making decisions and given the authority to enter into binding agreements on behalf of the company. Shareholders usually do not participate in running the business. In any case, they may not make commitments on behalf of the company simply because they are shareholders unless they hold a responsible office or position in the corporation. In other words, merely holding ten or one hundred or one thousand shares of General Motors does not give you authority to order a million dollars' worth of rubber tires from your friend's factory and have it shipped to the nearest Chevrolet plant.

One major difference, from a legal point of view, between the real person and the "artificial person" is that a real person, even acting in the role of a business person, is recognized by all states. So long as the real person meets certain minimal requirements for doing business—licenses, local taxes, etc.—he is free to come and go across state boundaries and to conduct his business accordingly. However, the legal "personhood" of a corporation is, strictly speaking, recognized only by the state in which that corporation was organized. A corporation established in one state may have to comply with certain legal requirements to do business in another state. While the general structure of a corporation is essentially the same from state to state, there are a number of differences among the state laws that must be observed. What these differences are and how they affect your business can best be determined by a good lawyer.

Legally, and practically, too, the life of a corporation is perpetual. Because the corporation is a separate legal entity, its existence is independent of the death or illness of any owner. Ownership is represented by the shares of stock and not by individuals. Furthermore, those shares of stock are property, which can be bought, sold, traded, and bequeathed.

There are other types of business structures—syndicates, trusts, joint stock companies, etc.—but these are designed to answer special needs of a highly sophisticated nature. Although they are too specialized to discuss here, your lawyer will know whether any of the less common business structures are right for you.

By the way, terms like "company," "associates," "firm,"

"organization," etc., have, at best, ambiguous legal or even commercial standing. "John Harris & Co." can be the name of a single proprietorship, a partnership, or a corporation. The "& Co." can be a work force of a thousand people or it can refer to Mrs. Harris, or John Harris, Jr. Even "John Harris" may designate someone who is dead, who has sold the business to Fred Smith, or who is, in reality, two people —John Brown and Harris Green. When a business is other than a single proprietorship, the company name often reflects the nature of the business structure—Smith & Jones Co., William White & Partners, The Stretch-O-Pod Corp., etc. Obviously, your company name cannot include words like "partners," "Corporation," "Incorporated," "Limited" (which is British for incorporated), or their abbreviations (Corp., Inc., Ltd.) unless your business conforms to those designations.

Money Matters

To determine how much money you will need to get started in business, it is necessary to decide how long you will have to operate before you can expect to see any returns. The answer to that question is largely a matter of your own good judgment and a knowledge of the business you are entering. As a general rule of thumb, assume that you will need to be self-sustaining for three months, although that can vary greatly with the product and the market. Relatively low-cost, rapid-turnover products should begin to bring profits very soon. Expensive items considered to be a major purchase by the consumer will take longer to show profits. You will have to make the necessary adjustments. Still, using our three-month rule-of-thumb, here are some of the typical factors to consider in determining how much money you will need to get started:

1. Salaries and wages. (The difference between "salaries" and "wages" is largely a semantic one. Usually, executives receive salaries, while laborers receive wages. Now that you know the difference, forget it; feel free to use the words interchangeably.)
2. Three months' rental of premises.
3. Three months' rental of fixtures and furniture.
4. Three months' leasing of equipment.
5. Promotion and advertising.
6. Three months' supply of raw materials.
7. Local, state, and federal taxes; licenses and other fees.
8. Office supplies: stationery, files and records, order forms, shipping documents, postage, etc.
9. Service charges: bookkeeping, accounting and legal fees, etc.
10. Financial costs: bank interest, bad debts.
11. Insurance: health, accident, fire, theft, etc.
12. Special factors and considerations.

By "special factors and considerations," we mean those aspects of setting up your business that are more or less unique to your business. If, for example, your product is a seasonal one, it may be necessary to manufacture during the off-season and have merchandise ready to ship at the beginning of the season. This will require storage space, which will have to be paid for.

As you can see by a quick glance at the above list, it is by no means universal or unshakable. You may decide to do business from your basement or your living room. That will certainly eliminate three months' rent for premises. It will also probably exclude costs for furniture. As for wages and salaries, if you and your spouse run the business yourselves, you may be willing to forego taking money out of it for a while. (But don't forget to include wages and salaries when calculating production costs.)

When you have finished estimating your financial needs

for getting started, take another look at the list. Make sure you have included everything that should be included. Be certain your estimates are fair and reasonable. If you err, let it be an error in overestimating rather than underestimating. Trim where you can so that excesses and little luxuries and conveniences are eliminated. (For example, if you decide to rent a typewriter, do you really need an electric one?) But there is a difference between trimming and skimping. Estimate only what you really need, but be generous in those estimates, or you may find yourself short of capital because of some unforeseen expense.

Once you know how much money you are going to need to get started, you will have to determine where you are going to get it.

Of course, you will do everything possible and necessary to make your business a successful one. But it is good business sense to realize, from the very beginning, that you are taking a risk. Like any sensible gambler, before you take the plunge, you must ask yourself: how much can I afford to lose? No doubt, your own resources consist primarily of savings. Most experts agree that everyone should have the equivalent of at least six months' income in a savings account to take care of any emergencies that might arise—such as unemployment or illness. Beyond that amount, only you (after consulting people whose opinion you respect and whose lives are likely to be affected by what you do with your savings) can decide whether you are prepared to risk the rest.

If your savings are insufficient to provide the capital you need, you may want to consider a bank loan. Again, the specific nature of your business will greatly influence the kind of loan you decide on. If your credit rating is good, a personal loan is probably the easiest to obtain. A business loan, however, is usually less expensive. Nonetheless, a business loan means you are inviting the bank to participate in your business and you will have to convince them that you and your enterprise are worth the risk.

Before you go to the banker for a business loan, do your homework. Come prepared with documents showing that you've worked out your break-even figures and your starting-up capital needs, and that you've made a careful market study. If you own stocks, bonds, mutual funds, or other investments, they can serve as good collateral for a loan, allowing you to borrow at a lower interest rate. When you use such securities as collateral, they continue to belong to you, and you continue to receive interest and dividends on them. You cannot, however, sell them while the loan is outstanding. Ownership is turned over to the bank only if you default on the loan.

(Speaking of investments, you may want to sell your stocks, bonds, or mutual funds, and invest the proceeds in your business. Doing so may give you more cash on hand than you could get from a bank loan, but it will also deplete your own resources. It may be better to borrow against an investment portfolio than to sell the contents of that portfolio.)

Another source of income is mortgages. First and second mortgages on your home or other real estate you might own are usually available.

It is often possible to pool your financial resources to avoid depleting any single reserve. For example, you may be able to raise the capital you need by taking a smaller mortgage and making up the difference with a portion of your savings. Or you may find it possible to supplement your savings by selling off, or borrowing against, only part of your investment portfolio. If you can work out such combinations, you may be able to retain some assets as a kind of cushion against unforeseen financial demands in the future.

Two points to remember when borrowing money from a bank are:

1) You will have to repay that loan, sooner or later, even if your business fails. Think carefully before you commit your future earnings, your investments, or your real estate. If you cannot repay the loan in any other way, you may have to give them up.

2) Money should be regarded as a commodity available for sale on the open market. As such, it can be had at varying prices, so shop around at several banks. One businessman I know avoided borrowing from banks in New York City, where he lives, and made arrangements instead with a bank in New Jersey because the New Jersey bank offered him better terms and a lower interest rate.

Once again, your lawyer or accountant can provide you with invaluable assistance in negotiating bank financing.

If you have any insurance policies, talk to your insurance agent about the possibility of borrowing against those policies. Arrangements vary from state to state, from company to company, and from policy to policy; it is impossible, therefore, to generalize. But your agent will be able to tell you what is available, at what cost, and at what risk.

Should you borrow from friends or relatives? That depends almost entirely on your personal relationships with those friends and relatives. A generous and doting father-in-law may be willing to bestow a few thousand dollars on you, fully realizing he may never see the money again. Of course, if you do lose the money he lends you, he may never let you live it down. Sometimes that is a much more difficult price to pay than simply repaying the loan. There is a good deal of emotion involved in borrowing from relatives and friends and, more often than not, these emotional factors interfere with the already less-than-smooth road to commercial success.

If you do borrow from friends or relatives, it may be to everyone's advantage to make the transaction a strictly business one. Signed notes, fixed rates of interest, and predetermined repayment dates help reduce some of the emotion to a practical business relationship.

What about outside investors? Many businesses obtain needed capital by inviting outsiders to participate. Investors participate in the risk, but they also participate in the return. Sometimes, when bank loans are not available, private investors will lend money to a businessman, usually at a rate considerably higher than the going bank rate. Unlike an "unlimited" or "silent" partner, such a lender usually does not share in the business's profits but receives, in addition to the principal he lent, interest at a predetermined rate. The principal and interest are payable whether the business succeeds or fails. To invite the involvement of such investors requires supreme confidence on the part of the new businessman.

Another way of encouraging or inviting outside investment is to form a corporation and sell shares in it. Remember, however, that each share is worth one vote and it is risky for a small businessman to sell more than fifty percent of the stock in his company. A stockholder invests in a corporation because he believes the company will make money and pay the shareholders a portion of the profits, called dividends. If that fails to happen, the shareholder has no recourse or legitimate complaint (so long as there is no evidence of mismanagement or fraud). You gamble; sometimes you win, sometimes you lose.

A popular way of attracting additional capital is to offer a partnership. In effect, the investor buys into your business by supplying part or all of the capital needed to get it going. There have been and are many highly successful partnerships. It is my personal opinion, however, that partnerships should be avoided whenever possible. As stated earlier, one characteristic of a general partnership is that any one partner can make commitments for the entire business. Probably more disputes and disagreements among partners have risen from that one fact than from any other.

Furthermore, differences of opinion arise over management, production, marketing, general policies, and every other aspect of running, maintaining, and expanding a business. When both partners have equal voices, dissension and rancor are almost certain to result. Again, I should emphasize that these are my personal views, based on my own contacts with small businesses. In keeping with that view, partnerships with close friends or with relatives should be avoided except in cases of dire necessity or abiding passion. Dire necessity means you have exhausted all other ways of obtaining the money you need. Abiding passion means you are so terribly fond of the person you want to take as a partner that you will be unhappy and unfulfilled unless the two of you can work together on your plans and dreams for the future.

Families and friendships have been irrevocably torn apart by business partnerships. If, therefore, you absolutely must go into business with a friend or relative, write down in precise detail what the responsibilities and obligations of each partner are to be, make sure that they are understood, and have the document signed, preferably before a notary public. So-called "gentlemen's agreements" deteriorate and collapse rapidly when one party to the agreement ceases to behave like a gentleman.

(Incidentally, the same general rule applies to employees: if you hire a friend or relative—or the son, daughter, niece, nephew, brother or sister of a friend or relative—you are begging for trouble.)

Going It Alone?

There are a few other factors, all highly variable according to the nature of the business and the product, that go into deciding the kind of business you are going to set up. These factors all revolve around whether you—that is, your business—will perform certain functions or whether you will engage the services of outside contractors to perform those functions for you.

1. *Can you do your own manufacturing?* The answer to that question depends, of course, on such considerations as the size, type, and cost of the manufacturing equipment required, the space available to house such equip-

ment, the capital required to pay for such equipment, and the operating personnel. In many cases, it may be possible to have a manufacturing firm produce your products under your label. That is exactly how supermarkets sell such a wide variety of food products carrying "house" brands, how drug stores sell vitamin pills with the drug stores' names on the label, and how liquor stores sell "private stock" wines and whiskies.

2. *Can you handle your own distribution?* It takes experience and time to understand and effectively utilize the various distribution and transportation channels available. Again depending on your personal resources—and that includes your knowledge, your capital, and the nature of the product and the business—it may be preferable to turn your entire production over to one or several wholesalers and distributors who not only have the personnel and experience to deal with such matters, but who can also distribute at lower costs because they combine your product with others they are handling for bulk shipments.

3. *Should you hire a salesman?* That depends on whether you can do your own selling or, if you cannot, whether you can afford to "carry" one or more salesmen on your payroll until they bring in enough sales to earn the money you are paying them. Even so, you may not be able to afford enough salesmen to cover all the territories you think are included in your potential market. In such cases, manufacturer's representatives may be the answer. A manufacturer's representative is an independent salesman who handles several related but noncompetitive product lines. A manufacturer's representative who already represents a line of cameras, another line of electronic flash units, still another line of lenses and filters, and a line of leather carrying bags, may be ideal for taking on one more line—tripods for seven-foot-tall amateur photographers. He is already calling on the retailers you are trying to reach, and since he

works on a commission basis, you have to pay him only for the orders he actually brings in.

4. *Should you go into the mail order business?* Many products lend themselves very well to mail order. There are two ways of selling by mail: The direct method and the indirect method. With the direct method, you advertise in various publications and invite potential customers to order the merchandise from you. You may also purchase mailing lists and send advertisements directly to your prospects. Direct mail advertising, however, may be surprisingly expensive and the returns may be surprisingly disappointing. It is better, therefore, to try a test mailing first. If you check the Yellow Pages under mailing list brokers, you are likely to find several near you, and they will be glad to discuss the kind of mailing list you should use and the kind of test you should run.

The indirect method of getting involved in the mail order business is to make your product available to companies that publish and mail catalogs offering a wide variety of items. Most companies of this type have very extensive mailing lists covering every area of the country. Check with your local librarian for a business directory listing mail order firms.

5. *Should you advertise?* The answer to that question is an unequivocal *Yes!* When calculating your start-up costs, you must include some budgeting for advertising and promotion. It may be small, it may be simple, but advertise you must. Remember what we said about the "better mousetrap" theory. No one will beat a path to your door if they don't know about your product. Whether you advertise through the mail, through print or air media, or by distributing handbills at a bus stop depends again on the nature of your business and product. What does not vary is that no one will know about your business if you don't tell them about it.

6. *Can you get publicity?* The essential difference between

advertising and publicity is that publicity is free. Because it is free, it is also somewhat less subject to your control. Nevertheless, it is well worth going after.

When it comes to publicity, think small. Maybe you and your invention would become overnight successes if you could have just twelve minutes on the *Today Show* or the *Tonight Show*. But those twelve minutes will be very hard to come by, and there is a lot of other free publicity around almost for the asking. Begin locally. The smaller the city or town you live in, the more eager the local media are for news and features about local people. Once you have received your patent, if you think your invention is truly unique, truly interesting to the general public, and one you would like to talk about, don't hesitate to call or write the city desks of the local newspapers, including semiweekly or weekly community papers. Nor should you hesitate to write or call the producers of local radio and television talk shows to tell them about your invention.

Most areas of the country are covered by one or more public television channels; that's the one that shows *Sesame Street*, performances of the Philharmonic, and innovations in modern dance. In recent years, these stations, in an effort to raise money, have been holding on-the-air auctions. By donating two or three units of your product to such an auction, you can obtain free television advertising that you couldn't buy with your entire advertising budget.

You have nothing to lose but a little time and postage, or the cost of a telephone call, by contacting the biggies. By all means, telephone or write to the editors of the major newspapers and magazines and the producers of network television and radio shows. They should not be overlooked. But neither should the local publicity outlets.

Where to Get Help

Earlier we mentioned that one mark of a good businessman is the ability to surround himself with experts who can advise him, guide him, and help him make decisions. There is a great deal of expert advice and assistance available to you, and most of it is free. Actually, it is free only in a manner of speaking. You, as a taxpayer and consumer, have already paid for these services to some extent; why not take advantage of them? They are there for you to use and, if you fail to seize the opportunity to use them, you may be taking an unnecessary risk with your business and your future.

1. *Federal Government.* In almost every field of business, the U.S. Department of Commerce has experts who get paid solely to guide and advise. Call the Department's nearest field office and make an appointment to meet with one of their consultants. When you get there, ask to see a list of available literature, most of which is either free or sold at nominal cost. You will be overwhelmed by the amount of available material that can help you with your business.

   The same holds true for the Small Business Administration, a Federal agency that also has field offices in many cities. In addition to advising you about your business and providing you with armsful of useful literature, the Small Business Administration can advise you on how to obtain a loan. The SBA may even be able to assist you directly in getting a small business loan.

   You will find the addresses of the Department of Commerce and SBA field offices in Appendix C.

2. *State Government.* About every state in the union has a Department of Commerce with offices in the state capital and major cities throughout the state. Check your telephone directory for the exact name of the department

in your state and its location. If you cannot find it, simply address an inquiry to "(*Fill in name of state*) Department of Commerce" and address it to the state capital city. If, for example, you live in Iowa and cannot locate the state agency by its actual name, address a letter to "Iowa State Department of Commerce, Des Moines, Iowa." Chances are excellent that your inquiry will get to the proper department.

There are also state and even city licensing agencies prepared to advise you on setting up your business in accordance with local laws. They may not be able to help you with such broad matters as policy and financing, but they can at least fill you in on fees, licenses, forms to be completed, etc.

3. *Private Agencies.* Throughout the country, there are various organizations prepared to help small businessmen, usually without cost. Most often, these organizations consist of retired executives whose combined know-how is worth billions of dollars. They are not hard to find; they often have "public service" announcements on radio and television, and state and civic organizations can usually guide you to them.

If the community in which you plan to set up business does not have a chamber of commerce, think seriously about locating elsewhere. The primary purpose of a chamber of commerce is the economic development of the area in which it exists, and if a community does not have one, it means that no one is particularly concerned with its development. The local chamber of commerce ("local" can sometimes encompass an entire state) is able, ready, and indeed anxious to help anyone who wants to go into business within its area. Depending on the chamber's own size and budget, it may help you find a suitable location, show you how to work out tax advantages and benefits, direct you to a good source of labor, help with distribution and transportation problems, etc. If nothing else, the chamber of commerce can thoroughly acquaint you with the business climate of the area you plan to work in. That is a very important thing to know.

It is inconceivable that you will be producing a product or entering a field of business not covered by one or more trade associations. Find out, either by checking directories at the library or by making inquiries at the Department of Commerce field office, the trade association covering your field of interest. Then make contact with that association. There is a wealth of information waiting to be tapped and brainpicking is encouraged in trade associations. Using this particular resource may require a small investment on your part: you may have to join the association or subscribe to its publication. In most cases, the cost is reasonable and well worth the investment.

4. *Trade Journals.* Your friendly librarian or the people at the Department of Commerce field office can acquaint you with trade journals in your field of interest. Chances are that neither the library nor the field office subscribes to the journal but they can give you the publication's name and address. In most cases, if you write and ask for a sample copy, you will get one, either free or for a small cost. You can then decide whether you want to invest in a subscription to the journal. In almost every case, the subscription is worth the price because of the journal's useful information.

I have saved for last the person who could prove to be your most valuable business "consultant"—the public librarian. No matter how small your local public library may be, it has two assets you should take advantage of. First, it is bound to have at least some of the directories and reference sources you will need. Second, and more important, it has a librarian—a trained, highly skilled person whose purpose in life is to help you find the reference material you need. If the librarian doesn't have it, he or she is likely to know where to get it. Even if you have nothing specific in mind, stop at the library during nonpeak hours, introduce yourself to the librarian in charge, explain that you are planning to go into business and expecting to need the library's resources, and ask for an introduction to those resources.

With all of these people ready and willing to help you, with a great deal of hard work, perseverance and study, and perhaps with a little bit of luck, you are ready to go into business yourself.

On the other hand, you may like your job, you may already be in a business that you don't want to give up, or you may feel that you simply are not the business type. What happens to your invention now?

Let's take a look at how you can make money from your patent in other ways.

CHAPTER

7

Selling Your Invention

In Chapter 5, we briefly touched on some of the ways an inventor can make money from a patent. We discussed the possibility of selling the patent outright for a flat sum of money, retaining ownership of the patent and allowing others to "practice" it, and, by far the most popular and probably the most practical method, selling the patent to a third party for a combination flat-fee-plus-royalty arrangement.

As mentioned, the Patent and Trademark Office will publish patents available for assignment in the *Official Gazette*. Companies in the market for new patents know about this and scan the *Official Gazette* regularly for likely prospects. Such companies, however, are few and far between. By all means, list your available patent in the *Official Gazette* on the chance that an interested party will notice it. You should also aggressively seek a market for your patent.

Finding the Market

Your time, energy, and money are precious. Don't waste them by trying to sell your products to people who have no use for them. Certainly General Motors, Westinghouse,

Chrysler, Ford, Monsanto, General Electric, IBM, and Lockheed are among the biggest corporations in the world. Yet, if you offer any one of these companies a tripod for seven-foot-tall amateur photographers, you may as well throw the inquiry letters down the nearest sewer as into a mailbox. There is the odd chance that a manufacturer of multiple-product lines might be interested. But it is exactly that—an odd chance.

Begin to look for a buyer for your patent in an intelligent and economical manner by zeroing in on your market. There are several business directories available, which list companies by category. You should have no problem finding the category covering your product, and it is the companies listed in that category who are your likeliest candidates.

Making Contacts

When you write to the companies, be brief. Give them a general idea of your invention and what it is supposed to achieve. If you do not yet have a patent, say so and be very general in your description. If you already have a patent, tell them and, if you want, be more specific. Letting the company know you have the patent will immediately make them more receptive to your idea.

One decision you must make is dictated to a great extent (like so many of your decisions) by the nature of your invention and the potential market. That decision is whether to approach large companies, medium-size companies, or small companies. There are advantages and disadvantages to each.

The primary advantage of a large company is, of course, its resources. Since it has the time, the money, and the personnel to develop your invention to a highly salable stage, it can do the necessary advertising and promotion to create a demand for the product. It can then fill that demand with an adequate, efficient sales and distribution setup. The chief disadvantage of dealing with large companies is that they're likely to be cool to new ideas coming from outside. A sizeable segment of a major corporation's annual budget goes to research and development (R&D). Scientists, engineers, and researchers are paid very high salaries to work in the R&D department and develop new inventions. Having already invested so heavily in inventiveness, a company is understandably reluctant to spend more of its money on something produced by an outsider probably unfamiliar with manufacturing processes and marketing problems. Furthermore, with all that inventing activity going on behind the corporation's closed doors, there is bound to be some duplication between what the company invents and what an independent outsider invents. By discouraging or severely restricting contributions from outside the corporation, the company avoids expensive and nasty legal arguments over allegedly "stolen" ideas and who invented what first.

Another disadvantage to having your invention picked up by a large company is that such a company has more than one product to promote. If yours does not produce successful results—that is, sales—immediately, the company may let your fish rot while it fries others. Of course, if the company buys your invention and it is a hit, you make more money with a major corporation than with a medium- or small-size company.

A middle-of-the-road company presents several advantages that the large company does not. For one thing, it may be possible to approach such a company on a somewhat less formal basis. (That formality will become clear to you a few pages from now when we discuss the specifics of how to submit an idea to a major company.) Instead of going through the R&D department, the production department, or the patent and trademark division, it may be possible to approach a sales manager or a marketing executive who has the imagination to recognize the potential of your product. He can then do the selling for you within the company. Because a medium-size company may not have the resources of a major corporation, it is less likely to have a highly developed R&D department. This means it could be more receptive to ideas from outsiders.

The disadvantages of a medium-size firm are, of course, the very things that are advantages for the large one. Resources for advertising, promotion, production, development, etc., are more limited.

A small company is the best in terms of doing away with the formalities. You can approach the president of the firm, very often in a telephone call or personal visit, explain what you have to offer (perhaps even show him the specifications, claims, and drawings of your patent), and discuss the marketing possibilities for your invention. The small company is likely to be even more receptive than the medium-size one to new ideas, precisely because it is small and does not have the R&D facilities of some of its larger competitors. A small firm probably has a much smaller line to sell and its sales efforts can concentrate more heavily on your product. But the main disadvantage of a small company is the exact opposite of the major corporation's greatest advantage: small companies have small resources. Even if most or all resources are spent on developing and promoting your invention, their total impact may be unimpressive.

Perhaps the best solution to the problem about what size company to approach is the buckshot method: Make your offer to all the companies, regardless of size, that you feel would be most likely to handle a product like the one you have invented. You can then decide whether to go with a large, medium, or small firm by the responses and reactions you get.

Let's take a closer look at a typical large company and consider some of the problems it must face in dealing with private inventors. Imagine, if you can, the number of letters a company like General Motors or Du Pont receives every year from crackpots who have invented a perpetual motion machine, a pill that will turn water into heating oil, a chemical that will grow hair on everything from a butternut squash to a wornout sofa, and so on, *ad infinitum*. Now, suppose that among those suggestions and ideas, there is one that makes a modicum of sense. Suppose, for example, some years ago someone sent a letter to General Motors claiming to have an idea that would enable automobile engines to use less gasoline. The method suggested may have been silly, even stupid. Nevertheless, General Motors, among other automobile manufacturers, has been working hard at developing engines that conserve fuel. Immediately on announcing such a development, however, GM may be subject to a lawsuit from that very crackpot, who may claim that GM "stole" his idea. It is necessary, therefore, for large companies to protect themselves against such lawsuits. Even when ideas come in from people who are not crackpots, there is, as mentioned earlier in this chapter, a very strong possibility that the company's R&D department is already working on an idea similar to the one the inventor is submitting. After all, if you devise some startling new improvement for typewriters, doesn't it make sense that a typewriter manufacturer with an R&D division working on ways to improve typewriters might duplicate or come close to duplicating your invention? In such cases, the prospect of a lawsuit looms.

To protect themselves from such contingencies, virtually every major corporation will return your idea to you, claiming it has been unread and not considered. In all likelihood, there will also be a form accompanying the returned letter and instructions specifying exactly how an idea is to be submitted. You will find that the form and instructions carefully spell out that the company is under no obligation to keep the idea secret, and that there is no confidential arrangement between the inventor and the company.

The 3M Company responds to all idea suggestions by mailing the inquirer a booklet entitled "3M Company and Your Idea." Bound into the booklet is an "Idea Submission

Agreement," which calls for: (1) a description of the idea, (2) an indication of its status—patent, patent application, unpublished manuscript, etc., (3) the name and address of your present employer and the employer for whom you were working when you first got the idea, (4) a place for the names and addresses of other people who may have rights to the idea, and (5) finally, a signed statement declaring that the person submitting the idea has read the booklet, "particularly the conditions set forth on Page 11," and agrees with each of the conditions contained in the booklet.

The "conditions set forth on Page 11" are:

(1) We will consider your idea only at your request.

(2) You must assure us that to the best of your knowledge you are the one who originated the idea, that you own the idea, that you have the legal right to negotiate with us concerning it.

(3) *We assume no obligation to you whatsoever in the absence of a written agreement entered into after we evaluate your idea.* [Emphasis added].

(4) *No confidential relationship between us is established when you submit your idea. 3M does not promise to keep it secret.* [Emphasis added].

(5) After we have studied your idea, we will tell you only whether we are interested in it. We assume no obligation to tell you about anything we knew previously or have discovered since you have submitted your idea.

(6) By submitting your idea to us, you are not giving us any rights under a patent or copyright you now have or may obtain in the future. On the other hand, 3M will not have any obligation to you in respect to any such patent or copyright, unless and until a formal written contract has been entered into between you and 3M.

Under these terms and conditions, if you remove 3M's name and insert the name of any major corporation you have in mind, you will have, essentially, the terms and conditions that most companies abide by.

To obtain a broad view of how corporations deal with outside inventors, I wrote to several large companies. Responses were received from (in alphabetical order) Bulova, Chrysler, Eastman Kodak, Exxon, General Electric, General Motors, IBM, Lockheed, and 3M. Since we have already seen 3M's approach to outside inventors, let's look at the others.

Bulova Watch Company, Inc., Bulova Park, Flushing, N.Y. 11370 wrote:

> With respect to "outside, independent inventors" we have the inventor sign a waiver of disclosure form [the form which acknowledges that there is no agreement of secrecy or confidentiality] and then evaluate the invention from technical and marketing viewpoints. If no patent has [been] issued on the invention and the person has no funds, we sometimes agree to prosecute the patent application for the inventor, but only if Bulova gets a license in return. On some occasions, an inventor assigns his invention to Bulova in return for monetary payment, but this is an extremely rare case with this company. More often, we take a license and agree to pay royalties only if a patent issues within X number of years.

Chrysler Corporation, Detroit, Michigan 48231:

> Chrysler has prepared a detailed pamphlet entitled "Patents, Inventions, Outside Suggestions, Trademarks and Copyrights." Although the booklet is intended for Chrysler personnel, they may let you have a copy if you ask for it. "Our corporation," Chrysler warns its em-

ployees, "is continuously confronted with the problem of how to safely consider unpatented suggestions which are made by parties outside of the corporation . . . The corporation has been exposed to a number of lawsuits and threats of lawsuits seeking payment of substantial sums which claims were based upon our use of structures that others had or alleged they had suggested to us. The defense of such lawsuits and claims, although heretofore quite sucessful, has involved considerable expense and has required the time and effort of numerous corporate employees which might otherwise have been more productively used."

To avoid such claims and lawsuits, Chrysler painstakingly spells out for its employees how to treat an idea that comes from outside the company.

Chrysler's employees are instructed to handle the correspondence gingerly. "The first step in the procedure followed by the Engineering Improvements Committee is to secure the suggestor's execution of our Suggestion Submission Agreement Form," Chrysler tells its employees. "This form is drafted to expressly avoid the assumption of any obligation by the corporation in connection with the consideration of the suggestion involved, except the obligation to respect valid patents." Once the signed Suggestion Submission Agreement Form is received, the idea goes through the necessary channels of technical evaluation and marketing considerations.

A copy of Chrysler's Suggestion Submission Agreement Form appears on Page 103.

Eastman Kodak Company, 343 State Street, Rochester, N.Y. 14650, wrote: "Kodak does receive many letters containing ideas and suggestions from outside the company *even though we do not encourage or solicit these disclosures.* [Emphasis added] While most ideas which are new to us and practical for our use originate within the company, we are always happy to receive proposals that others wish to bring to our attention." Kodak has a booklet entitled "Information on Submitting Ideas, Suggestions and Developments to Eastman Kodak Company and its Divisions and Subsidiaries," which describes the company's policies in reviewing these proposals. Kodak's terms are not unlike those already mentioned, and the booklet, of course, includes the usual form. An interesting variation in Kodak's conditions is Condition No. 4:

> Whenever the ideas, suggestions or developments submitted to the company are not protected by a valid patent or as a presently published copyrighted work, an honorarium will be given to the submitter if the submitted ideas are original with the submitter, are new and are adopted for use by the company as a result of the submittal. The amount of this honorarium will be established according to the company's reasonable judgment, but it will not in any event exceed $2,000.

In other words, if you do not have a patent, the most you can expect from Kodak is $2,000 if they decide to use your idea. Still, that's more than G.E. offers (see below).

Exxon Company U.S.A., P.O. Box 2180, Houston, Texas 77001, writes as follows:

> Exxon U.S.A. receives many invention disclosures from outside independent inventors each year. We recommend that such inventors file patent applications covering their inventions before submitting them to the company, but will accept disclosures from individuals who have not filed applications. All such inventions must be submitted to the company on a non-confidential basis and the inventor must agree to rely upon his patent rights to protect the invention. The compensation to be paid for an invention accepted by Exxon U.S.A. is a

CHRYSLER
CORPORATION

Suggestion Agreement

CHRYSLER CORPORATION
Outside Suggestions Department
CIMS 418-05-14
Post Office Box 1118
Detroit, Michigan 48231

Gentlemen:

I wish to submit a suggestion for the consideration of Chrysler Corporation which term I understand for the purposes of this Agreement also includes its domestic and foreign subsidiary, controlled, and associated companies.

My suggestion relates to:

I understand that you are willing to consider suggestions made to you but, because of the large number and the conditions under which such suggestions are submitted to you, you have established a uniform policy in regard thereto and that a statement of this policy is printed on the reverse side of this sheet. The policy requires that I accept the following specific conditions before consideration of my suggestion:

1 Chrysler Corporation is willing to consider any suggestion which may be submitted to it, but does so only at the request of the person making the suggestion. No suggestion information is received in secrecy or confidence regardless of any marking thereon to the contrary.

2 No obligation of any kind is assumed by, nor may be implied against Chrysler Corporation for any claimed or actual use by it of all or any part of the suggestion, notwithstanding any notice to the contrary on any information supplied to Chrysler Corporation. The only obligation which Chrysler Corporation shall have is that which is expressed in a formal written contract that may be executed by the parties after Chrysler Corporation has evaluated the information furnished under this Agreement.

3 I do not hereby give Chrysler Corporation any rights under any patents, trademark registrations, or copyright registrations I now have or may later obtain covering my suggestion, but I do hereby in consideration of the examination of my suggestion, release it from any liability in connection with my suggestion or liability because of use of any portion thereof, except such liability as may arise under valid patents, trademark registrations, or copyright registrations now or hereafter issued. However, if I should be an employee of Chrysler Corporation, nothing herein shall be deemed in any manner to alter or detract from any shop rights or the like which may be applicable for the benefit of Chrysler Corporation in regard to my suggestion.

4 In order that Chrysler Corporation may have a permanent record of all submissions made to it, I agree that there is no obligation to return any material submitted for its consideration, other than prototypes whose return is requested because of their value to the suggester.

I am agreeable to these conditions and ask you to consider under them my above-mentioned suggestion as well as any other suggestions I may hereafter submit to you.

Signature		Date	
Street address			
City	State		Zip Code

matter for negotiation between the company and the outside inventor. Such inventors can improve the chances that their inventions will be of interest to Exxon U.S.A. by working in technical areas which relate to the company's business, familiarizing themselves with what has been done in those technical areas in the past, developing their inventions sufficiently to make them practical, and considering the economic factors pertaining to their inventions.

General Electric Company, Fairfield, Connecticut 06431, deals with outside inventors through its Submitted Ideas Operation, located at the company's corporate headquarters.

All unsolicited disclosures received from outside the company are immediately passed to Submitted Ideas where they remain while Submitted Ideas personnel communicate with the submitter. If the idea is submitted gratuitously, the submitter is asked to sign an acknowledgement to that effect. If the submitter desires compensation, he is asked to sign a form covering the terms under which the company will consider his idea. In particular, these terms provide that if the submitter does not have or obtain a patent, any compensation will be limited to a maximum of $1,000. Also, the amount shall be payable only if the idea is new and is used [by G.E.]. These terms are essential since if there is no patent coverage, anyone else can freely copy the company's product, utilizing the idea at no cost whatsoever. Only after one of these two arrangements is agreed upon in writing is the description of the idea disseminated to the appropriate personnel in the company for business or technical evaluation.

Initial agreements reached with submitters are clearly indicated to be separate from any patent right that the submitter may have. Thus, the submitted idea is treated in a standard manner without regard to its patent status. We do not ask the submitter to release his patent rights in any way and, in fact, prefer to consider patented rather than unpatented ideas since the patents provide a clear, publicly available description of the concepts involved. Should the idea prove interesting to the company, patents are dealt with in a separate license negotiation where the submitter is free to ask for whatever compensation he feels appropriate for use of his patent rights.

In other words, General Electric, like Eastman Kodak, does not engage in a royalty or any other ongoing payment arrangement for a mere idea. Thus, if you approach a major corporation with a patent in your back pocket, not only is your invention protected, but you are negotiating for financial arrangements from a stronger position.

General Motors Corporation, New Devices Section, General Motors Technical Center, Warren, Michigan 48090: GM sent their booklet entitled "Submitting Ideas and Suggestions to General Motors." The pamphlet spells out the terms and conditions under which ideas are accepted for consideration; these are similar to those already given. Interestingly, however, GM made no mention of a waiver of disclosure, either in their correspondence or their booklet. But they do say that anyone who expects his idea to be kept secret should send it elsewhere because "none of our employees has the authority to accept any submission in confidence or to agree that any submission will be treated confidentially or in secret . . . If any submission is sent to us as confidential, it will not be considered. Instead, it will immediately be placed in a special locked file in the New Devices Section and kept there until we can determine what the sender wishes done with it."

International Business Machines Corporation (IBM).

Armonk, N.Y. 10504: IBM has a stimulating program that encourages inventions from its employees. But the company is so reluctant to deal with outside inventors that they specifically request that their letter on ideas from nonIBM personnel not be quoted and that the agreement they ask outside inventors to sign not be reproduced. If you want to sell your invention to IBM, you must communicate with them directly.

Lockheed Aircraft Corporation, Burbank, California 91520: If you have an invention that you think may be useful to Lockheed, write for a copy of their "Statement of Policies Concerning Submitted Suggestions, Ideas and Inventions." It looks and sounds a lot like the other companies' policies and includes a "Confidential Disclosure Waiver Agreement."

On the Inside

Each company listed above has a plan—some simple, some elaborate—for compensating employees who submit inventions to the company. Usually, if you work for a major corporation, anything you invent during work hours or using the company's facilities belongs to the company. If you invent something entirely on your own, in your own free time and away from the company premises, that invention may still be the property of your employer so long as you work for the employer. (You may want to review your employment file and see what agreements you signed when you were hired.) In most cases, however, there are substantial financial rewards for good ideas submitted to the company you work for. Check it out with your supervisor, your company's personnel department, or the employee relations department.

Where to Get Help

Just as there are agencies ready, willing, and able to help you if you decide to go into business, there are people prepared to assist you in selling or licensing your patent—often the same people. The field offices of the United States Department of Commerce, your state department of commerce, and similar regional agencies and organizations can help. A patent lawyer can also help find a customer for your invention.

Patent Developers and Brokers

As an inventor or a prospective inventor, your attention has surely been caught by the advertisements, large and small, of firms and individuals offering a variety of services. These services may include conducting a patent search and obtaining a patent, but almost certainly include finding a licensee for your patent. Be very careful in dealing with such companies. To begin with, if they advertise, that automatically means they are not patent attorneys or patent agents licensed to practice before the Patent and Trademark Office. Furthermore, there have been problems with some of these companies. Those problems can best be summed up by quoting from an article that appeared in the prestigious publication, *Science News*, of July 24, 1976:

> Worst of all . . . is the growing scandal over so-called "invention brokers" to whom the independents turn when they fail elsewhere.
> Some 250 such brokers were operating in the country last year [1975] dealing with roughly 1,000,000 inventors and doing an estimated $100 million of business. According to an investigator for the Federal Trade Com-

mission, the average customer of such brokers paid $1,000 to $1,500 for generally useless services and received almost nothing in return. When a California law recently required such brokers to disclose their records of success, one of the largest revealed that of 30,000 inventors it had contracts with, only three had earned a profit.

If you decide to deal with one of these "invention brokers," don't be impressed by the size or lavishness of their operations. I wrote two magazine articles about inventions, each of which featured a major invention broker, only to discover that after those articles were published, both brokers got into trouble with the Federal Trade Commission, the states' attorney generals or both.

Before you enter into any agreement with a patent broker, invention developer, or whatever other name the operator uses, check your nearest office of the Federal Trade Commission to find out what litigation or action is pending or has been completed against the broker. Also check with the local office of the Better Business Bureau. If the broker seems to have a clear record, ask some hard and important questions before doing business with him. Find out how many clients he has. Ask how many of the inventions he's handled have proven to be successful. (Don't be surprised if the answers you get to that question are vague and indeterminate. When I asked that question of a patent broker, the response was a lead-in to a philosophical discussion on how one measures success.) Ask for names, addresses, and telephone numbers of clients whose inventions have been successfully handled. If you are not satisfied with the answers you receive to any of these questions, run, do not walk, to the nearest exit. Don't be impressed by fancy booklets, slide presentations, slick magazines, or any of the other trappings so typical of patent developers. Results count; nothing else does.

CHAPTER

8

What Else is New: Designs, Trademarks and Copyrights

As was pointed out at the beginning of this book, not every new idea qualifies for one of the available types of patents. That does not mean, however, that all new ideas must be either abandoned or exposed to the mercy of ruthless and inconsiderate competition. There are three major methods of protecting new ideas—*design patents, trademark registration, copyrights*—and a fourth but minor one, which we have referred to briefly from time to time—*plant patents*.

Plant Patents

Plant patents protect "any distinct and new variety of plant, including cultivated sports [genetically, a mutation], mutants, hybrids, and newly found seedlings other than a tuber-propagated plant or a plant found in an uncultivated state," that has been asexually reproduced (i.e., reproduced by some means other than from seeds, such as by the rooting of cuttings, by layering, by budding, grafting, etc.).

The Patent and Trademark Office will be glad to supply more detailed information on obtaining a plant patent. (In 1970, the Plant Variety Protection Act was passed. This act provides a system of protection for sexually reproduced plants. Such plants come under the jurisdiction of the Department of Agriculture and information can be obtained by writing to the Commissioner, Plant Variety Protection Office, Consumer and Marketing Service, Grain Division, 6525 Bellcrest Road, Hyattsville, Maryland 20782.)

Design Patents

Let's take another look at our hypothetical venture, the collapsible, portable tripod for seven-foot-tall amateur photographers. The concept of a tripod is certainly not new. The idea of a tripod that can collapse for compactness is also not new. Makiing such a tripod from a sturdy, light-weight material may be new, but probably doesn't help us get our patent. This is because patents generally are not granted on the basis of the material used to manufacture the product. If we devise a truly unique and unusual method of manufacturing the tripod, we should look for a patent on the manufacturing technique and not on the end-product of that technique. For the moment, it looks as though there is no chance of patenting our tripod. But all is not lost.

It may turn out that what is different about our tripod is the way it is designed. In other words, if we can make our tripod look substantially different from the way other tripods look, we may be able to get a patent, anyway. This type of patent is called a *design patent*.

"The patent laws," advises the PTO, "provide for the granting of design patents to any person who has invented any new, original and ornamental design for an article of manufacture. The design patent protects only the appearance of the article and not its structure or utilitarian features."

Applying for a design patent requires essentially the same procedure as applying for a mechanical patent, except that it is usually both easier and cheaper. A search must be made to determine that the design, for which a patent is being sought, has not been previously patented or published. The application must include all the elements of a utility patent application, but the specification for a design patent is short. Only one claim is permitted. While a drawing is required and that drawing must conform to the PTO's standards for drawings, reference characters (the numerals to which various parts of the drawings refer) are not required.

The filing fee for a design patent application is $20. The issue fee depends on the term of the patent. A three-and-a-half-year term costs $10, a seven-year term is $20, and a fourteen-year term is $30. Once the PTO determines that the applicant is entitled to a design patent, it sends a "notice of allowance" to the applicant, his attorney or agent, and asks for payment of the issue fee to cover whatever term the applicant desires. (If the applicant is fairly certain at the time of filing how long he will want his patent to be enforced, he may include the issue fee with his application fee.) You'll find an example of an actual design patent on pages 137-139 in Appendix A.

Trademarks

We may also want to give our tripod some unique name, descriptive of its function. For example, we may call it "Stretch-O-Pod." If so, to ensure that no one else cashes in on our advertising and promotion, we must protect that name. To do so, we register it as a trademark.

The PTO defines a trademark as "a word, name, symbol, or device, or any combination of these, adopted and used by a manufacturer or merchant to identify his goods and distinguish them from those manufactured or sold by others." In other words, a trademark is a brand name used on merchandise "moving in the channels of trade."

The simple act of adopting a trademark and using it in the normal course of your business is enough to establish your rights to that trademark; it is not necessary to register the mark with the PTO. Once we begin to manufacture, distribute, and sell Stretch-O-Pods, should a competitor come along with a tripod called, say, "Stretch-A-Pod," we may obtain a court order against his continuing to use the name by proving that we were using it first. Still, rather than take a chance on something like that happening, and getting involved in expensive and time-consuming litigation, we are probably better off registering our trademark with the PTO. Registration, says the PTO, "constitutes the notice of the registrant's claim of ownership; and it creates certain presumptions of ownership, validity and exclusive right to use the mark on the goods recited on the registration."

The PTO maintains a *Principal Register* and a *Supplemental Register*. The *Principal Register* covers trademarks that are highly inventive or suggestive. Such marks, called "technical marks," are considered to be distinctive and more defensible. The *Supplemental Register* covers trademarks that consist of descriptions, names, geographic locations, and packages and labels, provided they have been in use for a year or more as a mark that distinguishes the applicant's products in ordinary trade. It is possible, after somewhat more use, to move a trademark from the *Supplemental Register* to the *Principal Register*. In any case, applications are made for the *Principal Register*.

Not far from the Patent Search Room at the PTO is the trademark examining facility of the Patent and Trademark Office, located in Crystal Plaza Building #2, 2011 Jefferson Davis Highway, Arlington, Virginia. Applicants who want to register a trademark should search the registered marks before filing their application. Here, too, professional searchers are available. In fact, chances are that a patent attorney or patent agent can also help you obtain a trademark registration. Of course, the PTO's records don't include trademarks that have been in use for either short or long periods of time if they have not been registered. Nevertheless, such marks do take precedence over the trademark being applied for.

To apply for a trademark, write to the Commissioner of Patents and Trademarks, Washington, D.C. 20231, and ask for a trademark application form, specifying whether the application will be filed in the name of an individual, a company or partnership, or a corporation. In the form, the applicant states that he is using the trademark for particular products, indicates the date when the mark was first used anywhere at all and the date on which it was first used in interstate commerce, and indicates how the trademark appears on the merchandise.

The applicant must also sign an affidavit stating that he believes himself to be the owner of the mark and that "to the best of his knowledge and belief, no other person, firm, corporation, or association has the right to use said mark in commerce, either in the identical form or in such resemblance thereto as to be likely, when applied to the goods of such other person, to cause confusion, or to cause mistake, or to deceive." The application, along with a drawing of the mark, five specimens or facsimiles of the trademark as actually used in commerce, and a $35 filing fee are returned to the PTO. If your application is rejected, you have six months in which to take action. This action may consist of a new application or an appeal to the Trademark Trial and Appeal Board, and, if necessary, to a court. If you fail to reply within six months, the PTO considers your application to be abandoned.

If your application for inclusion in the *Principal Register* is accepted, your trademark will be published in the PTO's *Official Gazette* for trademarks. That gives anyone who believes he has prior rights to the trademark, or who thinks that your use of the mark will be damaging to him in any way, an opportunity to raise objections. If no such objections are raised within thirty days, the trademark will be considered registered and the applicant will receive a Certificate of Registration.

Trademarks accepted for the *Supplemental Register* are also published in the *Official Gazette* for trademarks, but without being subject to any objections.

A trademark is registered for twenty years from the date of issue. At the end of each twenty-year term, the registration may be renewed. This may continue for an indefinite number of terms, so long as the trademark is still used in commerce.

Many people confuse trademarks with trade names. Trade or commercial names are business names used by companies to identify their businesses. You may not register a trade name. On the other hand, if your trade name includes a registered trademark, no one else may adopt your trade name since that would infringe on the registration of the trademark. In other words, if you set yourself up in business as Smith Tripod Manufacturing Co., and, in another part of the country, another Smith goes into the tripod business and calls himself the Smith Tripod Manufacturing Co., you have little, if any, recourse. But if you call your company the Stretch-O-Pod Manufacturing Company, and Stretch-O-Pod is a registered trademark, no one else may use that name. (There are often state and local laws regarding duplications of company names. If the other Smith decides to go into business in your city, county, or state, he may be prevented from using a name under which you are already doing business.)

A *service mark*, however, may be registered. Obviously, not every company is engaged in producing products. A distinctive name describing a company's service is the equivalent of a brand name for merchandise. The Associated Press, for example, provides a service rather than a product and is entitled to protection from other news-disseminating agencies also wanting to call themselves the Associated Press. Service companies such as airlines, automobile rental companies, insurance companies, banks, etc., may register service marks that uniquely and distinctively describe the services they offer.

Not all trademarks are eligible for registration. As we stated earlier, you may not register a trademark that someone else has been using for years or that is blatantly deceptive. (It is doubtful, for example, that the PTO would grant a trademark for brass vessels that an applicant wants to call "Pot O'Gold.") Nor may you register a trademark that, in the eyes of the PTO, is immoral or that includes the American flag or any of the insignia of the United States.

There is one exception to using a trademark that someone else already owns. It may be possible to obtain trademark registration if the product to which the trademark applies is completely unrelated to the product or service of the previous trademark owner. To offer a farfetched example, we should have no trouble registering Stretch-O-Pod for tripods even if an agricultural supply house has registered the same name for a chemical fertilizer guaranteed to give a higher yield to green pea farmers.

When choosing a trademark, try to keep it short, simple, and descriptive. Be sure to avoid obscenities or swear words in foreign languages.

The time may come when you decide to abandon your trademark. Such an act is usually taken with mixed feelings. On the one hand, it is gratifying to know that your product has become so universally known and accepted that its brand name is applied to all products of its kind. On the other hand, it is a little disheartening to know that you have worked so hard to achieve recognition that others are enjoying. Words like cellophane, aspirin, harmonica, linoleum, nylon, and rayon were once registered trademarks. Trademarks like Xerox, Kleenex, Vaseline, and Band-Aids are having a hard time trying to avoid joining the ranks of those product names that are now generic (i.e., generally descriptive words).

Like patents, a U.S. trademark registration offers no pro-

(FORM FOR USE OF INDIVIDUAL)
(Instructions on reverse side)

U.S. DEPARTMENT OF COMMERCE
PATENT AND TRADEMARK OFFICE

APPLICATION FOR TRADEMARK/SERVICE MARK REGISTRATION
(DECLARATION)

Mark _____
(Identify Mark)

Class No. _____
(Insert number, if known)

TO THE COMMISSIONER OF PATENTS AND TRADEMARKS:

1

2

3

[4] a citizen of

The above identified applicant has adopted and is using the mark shown in the accompanying drawing for[5]

and requests that said mark be registered in the United States Patent and Trademark Office
on the Principal Register established by the Act of July 5, 1946.

The mark was first used on[6]   , 19   ; was first used
in[7]   commerce on[8]   , 19   ; and is now in use in such
commerce.

The mark is used by applying it[9]

and five specimens showing the mark as actually used are presented herewith.

The undersigned applicant [10] _____ declares:
That he believes himself to be the owner of the mark sought to be registered; that to the
best of his knowledge and belief no other person, firm, corporation, or association has the
right to use said mark in commerce, either in the identical form or in such near resemblance
thereto as may be likely, when applied to the goods of such other person, to cause
confusion, or to cause mistake, or to deceive; that all statements made herein of his own
knowledge are true and that all statements made on information and belief are believed to
be true; and further that these statements were made with the knowledge that willful false
statements and the like so made are punishable by fine or imprisonment, or both, under
section 1001 of Title 18 of the United States Code and that such willful false statements
may jeopardize the validity of the application or document or any registration resulting
therefrom.

11

_____
(Date)

_____
(Signature of applicant)

[Enclose Filing Fee of Thirty-Five Dollars]

INSTRUCTIONS

1. Insert name of applicant and if applicant has a trade style or name, insert "doing business as _____" following the name.

2. Insert the business address: street, city, State, and ZIP Code.

3. Insert residence address: street, city, State, and ZIP Code.

4. Insert country of citizenship.

5. Name by their common, usual, or ordinary commercial name (e.g., canned fruit and vegetables) the products or goods on which the mark has actually been used.

6. Insert the date of first use anywhere on *any of the goods* recited in the application.

7. Insert the kind of commerce, i.e., "interstate" or "Territorial" (District of Columbia, Virgin Islands, Puerto Rico), or such other specified type of commerce as may be regulated by Congress. Foreign applicants must specify: "commerce with the United States."

8. Insert the date of first use on any of the goods in interstate or foreign or Territorial commerce, as the case may be.

9. Insert the method of using the mark on the goods, i.e., "to containers," "to labels applied to containers," "to tags or labels affixed to the goods," "to name plates attached to the goods," "by stamping it on the goods," or other appropriate method, and send specimen labels, or containers, or name plates, etc., showing such use. If the specimens are third dimensional, photographs should be supplied in lieu of the article itself, except if the specimen is a carton or box which can be folded to a size not exceeding 8½ by 13 inches.

10. Insert name of individual executing declaration.

11. Individual's signature.

NOTE

If an attorney at law, or other person who is recognized by the Patent and Trademark Office in trademark matters, is to file and prosecute this application. a *signed* power or authorization in substantially the following form should accompany the application:

Please recognize _____ , *

_____ , †

with offices at _____ ,
to prosecute this application, to transact all
business in connection therewith, and to receive
the certicicate.

---

*Insert the name of the individual attorney at law or the law firm, or other recognized individual or firm, as the case may be.

†If an individual attorney at law, insert "a member of the bar of the State of" and name the State of admission to the bar.
  If a law firm, insert "a firm composed of" and list the names of the members of the firm and their States of admission.
  If a recognized individual other than an attorney at law, insert Patent and Trademark Office Registration number.
  If a firm of nonlawyers, insert "a firm composed of" and list the members of the firm and the Patent and Trademark Office Registration number of each. Names of firms of nonlawyers may not be used unless all members of such firm are recognized under Rule 2.12(b) of the Rules effective August 15, 1955.

NOTE

The drawing must be made upon pure white durable paper, the surface of which must be calendered and smooth. India ink alone must be used for pen drawings to secure perfectly black solid lines.

The size of the sheet on which a drawing is made must be 8 to 8½ inches wide and 11 inches long.

A heading should be placed on the left-hand side at the top of the drawing. This heading consists of the applicant's name, his post office address, the first dates of use, and the goods.

If the application is for registration only of a word, letter, or numeral, or any combination thereof, not depicted in special form, the drawing may be the mark typed in capital letters on paper, otherwise complying with the requirements.

tection in foreign countries. A U.S. registration may, however, serve as a basis for obtaining similar registration in several countries.

Reproduced on pages 116-117 is an actual application form for a trademark and service mark.

Copyrights

Copyrights have nothing to do with inventions; they come under the jurisdiction of the Library of Congress.

The purpose of a copyright is to protect the copyrightholder's writings against copying. A copyright does not protect the subject matter but only the manner in which it is expressed. The book you are now reading is an example of this principle. It is conceivable that another author, on reading this book, may decide he would like to write a book on how to patent and market your invention. There is nothing in the world to prevent him from doing so. But because this book has been copyrighted, no one else may use the same words, in the same order, as they appear here. (Except for material taken from government publications; such publications are not covered by copyright except when quoting previously-copyrighted material; it gets a little complicated sometimes.)

Let's go back to our tripod for seven-foot-tall amateur photographers—which we can now begin referring to as the Stretch-O-Pod. If we write out a careful description of the Stretch-O-Pod and have that description copyrighted, what we have protected is *the description,* not the product being described. Thus, a copyright can be a valuable form of protection for an inventor who, unable to obtain a patent, may want to sell plans or instructions for creating the device he has invented. But it cannot protect the invention itself.

Copyrights are issued for fourteen classifications of original work. These include: books, periodicals, speeches, plays, musical compositions, maps, artwork, reproductions of art-work, technical drawings and models, photographs, commercial prints (such as greeting cards), motion pictures, documentary-type films, and sound recordings.

For detailed information about copyrights, write to: Register of Copyrights, Library of Congress, Washington D.C. 20540.

CHAPTER

9

Checklist: Your Invention from Start to Finish

This chapter will serve as a handy checklist for you. Almost everything given here, in a sentence or two, is discussed in detail in the preceding chapters. Make sure you fully understand and agree with all the items shown. If you don't, go back and review the material.

Getting the Patent

☐ Invent something that is both practical and marketable. There are, among the more than four million patents issued by the Patent and Trademark Office, more than a few "vanity" patents. Such patents, which are harmless of course, are also unrewarding from a financial point of view.

☐ Don't talk or write about your invention. If you expect to receive a patent, your invention must be novel and not previously known.

☐ Keep careful records from the very beginning. Write your

idea in a notebook as soon as it occurs to you. Also begin recording the development and refinement of your invention. Be sure your records are witnessed periodically.

☐ Have a search made of "prior art"—patents already in existence and published literature on the product you have invented. Become familiar with patents, especially those in the classification and subclassification of your own invention. Get used to the language of patents and the drawing style.

☐ Don't reveal the date on which you filed your application or the serial number assigned to it by the PTO while your patent application is pending at the patent office.

When you obtain your patent, you must decide whether to use your invention as a basis for your own business, or whether to sell the patent in exchange for financial reward.

Going Into Business

If you decide to go into business for yourself, check the following:

☐ Determine, as objectively as possible, whether you are the type of person to establish and operate a successful business. This is most important.

☐ Examine and study carefully, to the extent of your knowledge, experience, and available funds, your potential market.

☐ Decide the kind of business you will establish—a single proprietorship, a partnership, or corporation—based on the nature of your product, the nature of your market, the available finances, and the amount of time and energy you are prepared to devote to that business.

☐ Do a break-even analysis to determine the minimum amount of business you must do to at least keep afloat.

Ask a bank executive, an accountant, or a counselor at the Department of Commerce or Small Business Administration to help formulate a break-even analysis for your prospective business.

☐ Estimate, as accurately as possible, the amount of money you will need to start your business and keep it going for at least three months.

☐ Establish your business with a view toward the future. Consider the constancy and continuity of your present and future markets, and the changes, adaptations, and innovations that, depending on how they are handled, can interfere with or greatly enhance the future growth of your business.

Selling Your Invention

If you decide to let someone else take over your invention, here are the points you should keep in mind:

☐ Take the time to seek out the likeliest market for your invention. Don't try to sell a gardening tool to General Motors or a tripod for seven-foot-tall amateur photographers to Monsanto Chemical.

☐ Write to the companies you think will be interested in your invention. Give only the broadest description of your idea, concentrating on what it is expected to accomplish rather than how it accomplishes it. Ask what the proper procedure is for submitting ideas to the company.

☐ Make sure you understand any agreement completely before signing it. If you don't, ask the company or your lawyer to explain it. (This is especially true for any agreement involving financial compensation.)

☐ Make sure you fully understand what rights you are giving up when you assign your invention to a company.

☐ Check out the company you work for. Many firms, both large and small, have programs, ranging from stingy to generous, covering inventions by their own employees.

Some Practical Advice

☐ Get as much professional help as your budget will allow. In the long run, professionals can probably save you money. For example, unless your patent application is an exceptionally simple one, you're likely to find that a registered practitioner (patent attorney or patent agent) can save you time, money, and mental anguish in conducting the preliminary search and preparing the patent application.

☐ Remember that registered practitioners are not permitted to advertise. Therefore, people who do advertise their services are, by definition, not registered practitioners. Also remember that many of those who do advertise have been or are in trouble with the Federal Trade Commission or their states' Attorney Generals' offices. If you decide to consider the services of a patent developer, check with the Better Business Bureau, the Federal Trade Commission, and the Attorney General's office in your own state.

☐ When you receive a patent, that fact is published in the *Official Gazette,* which is, of course, a public document. Shortly thereafter, you are likely to hear, either by mail or telephone, from companies or individuals offering to act as your agent in selling or licensing your patent. What has been said above for patent developers holds true for these agents. Check them out carefully.

It should be noted that not all patent developers or brokers are suspect. According to what one prominent international patent attorney told me, some of them are disbarred registered practitioners. There are, however, a number of patent developers and brokers who are highly successful and highly reputable. The inventor must decide first whether he needs the help of such an organization or individual and, second, whether, on the basis of his own research, he has determined that the organization or person is reputable.

☐ Get all the free help, in addition to the professional help recommended above, you can. The U.S. Department of Commerce, The Small Business Administration, state departments of commerce, chambers of commerce, regional and local development organizations, banks, and trade associations are just a few of the many types of organizations ready, willing, and able to help, often for little or no cost.

☐ Check with the U.S. Department of Commerce and the Small Business Administration for long lists of booklets and pamphlets. These contain both general and specific business information.

The following books and pamphlets relate directly to patents and trademarks and are available as indicated:

The U.S. Patent and Trademark Office, Washington, D.C. 20230, will send, on request, single copies of the following pamphlets at no charge: *Q & A About Patents; Q & A About Plant Patents; Q & A About Trademarks; General Information Concerning Patents; General Information Concerning Trademarks.*

The following material may be ordered from the Superintendent of Documents, U.S. Government Printing Office, Washington, D.C. 20402 (prices shown were current at the time of writing and are subject to change): *Patents and Inventions: An Information Aid for Inventors,* 50¢; *Patent Laws,* $2.10; *Attorneys and Agents Registered to Practice Before the U.S. Patent and Trademark Office,* $3.70; *Code of Federal Regulations: 37, Patents, Trademarks and Copyrights,* $2.20; *Guide for Patent Draftsmen,* 65¢.

In addition, there are thick, technical, loose-leaf-bound manuals on examining procedures, classifications, and indexes. Subscriptions are also available to the patent *Official Gazette* and the trademark *Official Gazette.* Although all are expensive and probably of only limited value to the individual inventor, if you want to know more about them, write to the Superintendent of Documents at the above address for more information.

CHAPTER

10

Your Future as an Inventor

It is worth repeating the quotation of Henry L. Ellsworth, the one-time Commissioner of Patents who said, in 1843: "The advancement of the arts, from year to year, taxes our credulity and seems to presage the arrival of that period when human improvement must end." There is apparently no record of· whether Commissioner Ellsworth had a specific date in mind, but there seems little danger that the end is imminent.

It seems that almost every year, innovations bordering on the miraculous appear. Man is creating life in a laboratory test tube. The prospects for the future from that one accomplishment alone stretch the imagination to the limits and are, indeed, a little frightening.

The 1960's and 1970's have seen developments in several major areas that have literally changed the course of human history. There can be no doubt that the human condition has been—and will continue to be—permanently changed by computer technology, for example. The Concorde SST has made air travel between the United States and Europe a morning's diversion. We have seen fantastic innovations in communication. A live telecast of a man walking on the moon probably draws less of an audience than a typical cops-and-robbers program. Indeed, in the same century in which telephones were at one time considered a luxury, color television is commonplace, and great historical events are witnessed by ordinary people all over the world.

Silicon chips, transistors, microwave ovens, instant photography, laser technology—all have contributed greatly to the health, education, and welfare of the human race. And each new technological achievement appears, not as the culmination or conclusion of development, but rather as a door opening on a long road of progress waiting to be explored.

Along with the earth-shaking innovations that the scientific community keeps presenting to us are the seemingly small, relatively unimportant, but nevertheless worthwhile inventions that make life a little easier. These include: Scotch tape and instant glue; nonreflecting picture frames, wrinkle-resistant clothes, nonbreakable phonograph records and toys; adjustable shower heads, electric knives, automatic garage-door openers, and pop-top beer cans. The list goes on and on and consists to a great extent of products, designs, and innovations that have appeared only during the past ten to twenty years.

Thus, if you do build that better mousetrap, perhaps the world will not beat a path to your door, but it will be ready to hear about it, to buy it, and to compensate you for your inventiveness.

Most inventions are made for and by employees of large institutions—corporations, universities, foundations. But there is still a hard core of working inventors, many of whom rely on inventing as their sole source of income. If they can do it, it's certainly possible that you can too.

Inventors are often depicted as dreamers and drifters, but a successful inventor is nothing if not practical. He knows that to succeed as an inventor, pep talks are not enough; he must deal with realities. So must you. If you want to consider the possibility of a future as an inventor, begin by taking these practical steps:

Keep Your Eyes and Ears Open

Be alert to developments and shortcomings in the areas you know best. Whether you work on a factory assembly line or in a chemical research laboratory, on a loading dock or in an office, you are probably constantly hearing complaints about some aspect of the work that is unsatisfying to the workers. Your best chances of success as an inventor may lie in those complaints.

To begin with, your coworkers, quite unknowingly, can provide a constant source of inspiration. Furthermore, if you can demonstrate to your employer an effective way to eliminate those complaints, you are well on your way to achieving some degree of success, if not as an inventor, then certainly as an innovator.

Study the Field

Keep abreast of developments in your field of interest. In addition to your job, a hobby can also provide considerable inspiration for an inventive mind. Photography, woodworking, astronomy, model-making, do-it-yourself home repairs and maintenance, and virtually any other avocation involving tools or equipment can provide you with inventive inspiration.

Broaden Your Scope

Also take a close look at the world beyond your immediate range. Energy conservation, electronics in general and computer technology in particular, aerospace, and environmental protection are among the areas of greatest interest right now. A device or a process in any of these areas might lead to success and riches.

When you do look beyond your own private world, don't overlook your immediate neighborhood—in other words, the average citizen-consumer. There are countless useful devices waiting to be invented and bought by the consumer market or those who serve it. For example, I want someone to invent self-adjusting theatre and airplane seats, bread wrappers that close tightly by themselves, child-proof caps for medicine bottles that are not only truly child-proof but also easily opened by an adult, smear-proof eyeglasses, and a perpetual light bulb.

Our world is a tough and competitive one, but it always shows respect for a person with guts, integrity, vision, and common sense. In the words of the sign that hangs over the entrance to the Public Search Room at the Patent and Trademark Office:

"Don't just stand there—invent!"

APPENDIX A

The Language and Art of Patents

Reproduced here is an actual utility patent for an electronic signal-delay device. The inventor, Dr. Karl F. Ross, is a prominent international patent attorney with offices in New York City and correspondents throughout the world.

This illustration not only shows an actual grant of Letters Patent, but also provides an excellent example of the language and phrasing to be used when preparing a patent application. Note particularly how each succeeding claim refers to one or more of the preceding claims.

This patent also offers a good example of what constitutes acceptable drawings for inclusion with a patent application.

Reproduced on pages 134-136 is another example of a utility patent. This one is for a Scouring Pad Containing Dispensable Substance, invented by Ada Bottino.

Reproduced on pages 137-139 is an example of a design patent for a television receiver designed by Melvin H. Boldt and Richard K. Althans.

Nº

2,998,543

THE UNITED STATES OF AMERICA

TO ALL TO WHOM THESE PRESENTS SHALL COME:

Whereas   Karl F. Ross,

of

Bronx,   New York,

PRESENTED TO THE **Commissioner of Patents** A PETITION PRAYING FOR THE GRANT OF LETTERS PATENT FOR AN ALLEGED NEW AND USEFUL INVENTION THE TITLE AND A DESCRIPTION OF WHICH ARE CONTAINED IN THE SPECIFICATION OF WHICH A COPY IS HEREUNTO ANNEXED AND MADE A PART HEREOF, AND COMPLIED WITH THE VARIOUS REQUIREMENTS OF LAW IN SUCH CASES MADE AND PROVIDED, AND

Whereas UPON DUE EXAMINATION MADE THE SAID CLAIMANT   is ADJUDGED TO BE JUSTLY ENTITLED TO A PATENT UNDER THE LAW.

NOW THEREFORE THESE **Letters Patent** ARE TO GRANT UNTO THE SAID

Karl F. Ross, his heirs

OR ASSIGNS

R THE TERM OF SEVENTEEN YEARS FROM THE DATE OF THIS GRANT

GHT TO EXCLUDE OTHERS FROM MAKING, USING OR SELLING THE SAID INVEN- OUGHOUT THE UNITED STATES.

In testimony whereof, I have hereunto set my hand and caused the seal of the Patent Office to be affixed at the City of Washington this twenty-ninth *day of* August, *in the year of our Lord one thousand nine hundred and* sixty-one, *and of the Independence of the United States of America the one hundred and* eighty-sixth.

Attest:

Ernest S. Swider
Attesting Officer.

David L. Ladd
Commissioner of Patents.

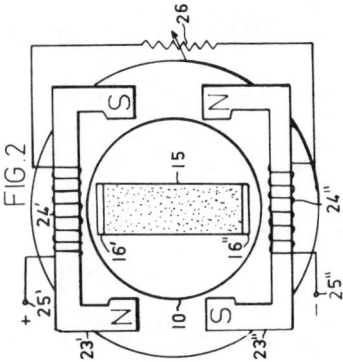

FIG.2

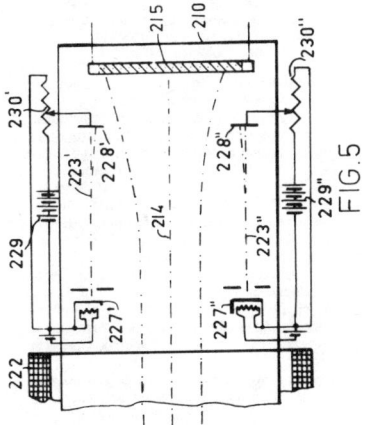

FIG.5

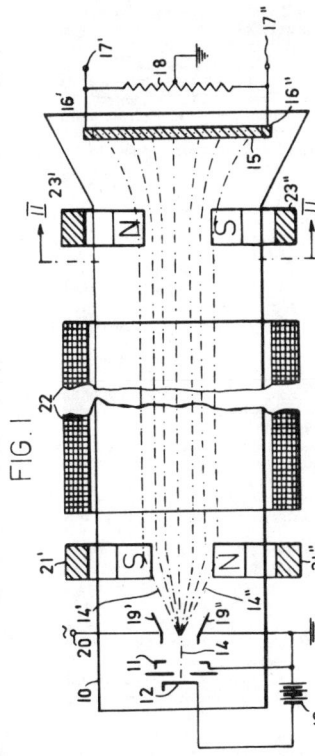

FIG.I

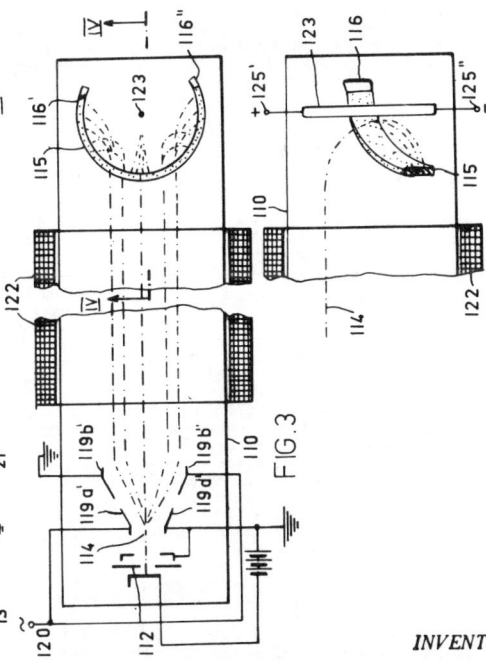

FIG.3

FIG.4

INVENTOR:

Karl F. Ross

1

2,998,543
ELECTRONIC SIGNAL-DELAY DEVICE
Karl F. Ross, 5121 Post Road, Bronx, N.Y.
Filed Jan. 6, 1959, Ser. No. 785,221
10 Claims. (Cl. 315—27)

My present invention relates to a device for producing a delayed output in response to a control signal.

For high-frequency signaling, in which even short absolute delay times correspond to substantial fractions of an oscillatory cycle, it is often convenient to use cathode-ray tubes as delay devices. This presents no particular problem if the signal to be delayed is applied to an intensity-control electrode of the tube so as to modulate the intensity of the electron beam thereof, this beam then impinging upon a suitable target electrode to deliver an output current proportional to said intensity. In many instances, however, better amplification or other desirable effects may be obtained by using the signal not to modulate the intensity of the beam but to control its deflection; in these cases the tubes become bulky and unwieldy if a large beam sweep is to be combined with a considerable phase delay.

The general object of my present invention is to provide an improved delay device of the character referred to in which the radial dimensions of a cathode-ray tube are maintained small over the major portion of its length, irrespectively of the magnitude of the delay introduced.

The invention realizes this object by the provision of means at the input end of a cathode-ray tube for deflecting a beam substantially parallel to itself, e.g. by two pairs of cross-connected electrodes or by magnetic means, in combination with magnetic focusing means maintaining the beam axially oriented throughout the major portion of the elongated tube envelope. Advantageously, in order to afford increased deflection without material enlargement of the tube, additional deflecting means may be provided at the output end thereof (i.e. near the target electrode) to convert the various parallel beam paths into divergent ones.

The invention will be more fully described with reference to the accompanying drawing in which:

FIG. 1 shows, somewhat diagrammatically, a cathode-ray tube embodying the invention adapted to serve as a signal-delay device;

FIG. 2 is a section taken on the line 2—2 of FIG. 1;

FIG. 3 is a view similar to FIG. 1, showing a modified delay device embodying the invention;

FIG. 4 is a section taken on the line 4—4 of FIG. 3; and

FIG. 5 is a view similar to the right-hand portion of FIG. 1, illustrating a further modification.

In FIGS. 1 and 2 I have shown a cathode-ray tube 10 whose evacuated cylindrical envelope contains an electron gun including an accelerating anode 11 biased positively, relatively to a cathode 12, by means of a battery 13. This gun emits an electron beam 14 toward a target electrode consisting of a body 15 of resistance material bounded by a pair of conductive terminal strips 16', 16", the latter being tied to respective output terminals 17', 17" connected across an output resistor 18 with grounded center tap. Since the anode 11 is likewise grounded, the region between it and target electrode 15 represents a drift space in which the electrons of beam 14 undergo substantially no further acceleration.

A pair of deflecting electrodes 19', 19", connected between ground and a signal-input terminal 20, serve to deflect the beam 14 from the tube axis into a variety of positions of which some, including the extreme beam position 14' and 14", have been indicated in dot-dash lines in FIG. 1. The beam then passes between a pair of horseshoe magnets 21', 21", which may be of the

2

permanent type, positioned on either side of the tube axis with relatively inverted polarity so as to produce a field which has a maximum intensity of one sign in the region of beam position 14', goes through zero at the tube axis and has a maximum intensity of the opposite sign in the region of beam position 14".

With proper selection of the dimensions, intensity and relative spacing of magnets 21', 21" it is possible to obtain a field whose strength in the median plane of tube 10 varies nearly linearly with the distance from the tube axis whereby, with suitable choice of polarity as indicated on the drawing, the electron beam deflected by electrodes 19', 19" will again be deflected into a substantially axial position irrespectively of the original angle of deflection. The thus re-oriented beam enters the axially extending magnetic field of a conventional focusing coil 22 serving to prevent the electrons from straying from their axially directed path. The length of this part of tube 10 is assumed to be such that the transit time of the electrons between input electrodes 19', 19" and output electrode 15 is equal to at least a substantial fraction of a cycle of the high-frequency signal fed in at terminal 20 whereby a desired degree of delay will be obtained.

Since the distance of the electron beam from the tube axis in the region of coil 22 will be a function of the input voltage applied across deflecting electrodes 19', 19", a useful output could be obtained with target electrode 15 positioned immediately at the end of this coil or even within the magnetic field thereof. In accordance with another feature of the invention, however, I prefer to provide a second pair of horseshoe magnets 23', 23" between coil 22 and target 15, the polarity of these magnets being opposite to that of magnets 21', 21" whereby the beam will be deflected outwardly, by an angle at least roughly proportional to the spacing of the beam from the tube axis, so that the loci of impingement of the electrons upon target 15 will be spread further apart and a larger target electrode, with improved signal-to-noise ratio, can be used.

The amplitude of the delayed output signal at terminals 17', 17" may be varied, according to another feature of the invention, by modifying the extent of the deflection of the beam under the control of magnets 23', 23". For this purpose I have shown each of these magnets provided with an energizing winding 24', 24", respectively, connected across a source of current (not shown) by way of terminals 25', 25" in series with an adjustable impedance indicated schematically at 26.

FIGS. 3 and 4 show a device somewhat similar to that of FIGS. 1 and 2, comprising a tube 110 from whose cathode 112 emanates a beam 114. The position of this beam is controlled by two pairs of deflecting electrodes 119a', 119a" and 119b', 119b" which are cross-connected between ground and a signal input terminal 120 in such manner as to maintain the deflected beam parallel to the axis of the tube, as disclosed in my U.S. Patent No. 2,728,894 issued December 27, 1955. The elongated central portion of the tube is surrounded by a focusing coil 122.

The output end of tube 110 comprises an elongated conductor 123 positioned at right angles to the tube axis and energized from a source of direct current (not shown) via terminals 125', 125"; the intensity of this current may be controlled by suitable means such as the variable impedance 26 of FIG. 2. Conductor 123 is surrounded by a cylindrical, non-homogeneous magnetic field whose strength perpendicular to the median plane defined by the conductor and the beam axis, which is transverse to the plane of deflection of beam 114, is a maximum in said median plane and tapers off rather sharply on either side thereof, both because of the increasing distance from conductor 123 and by reason of the smaller angle en-

3

closed between the circular lines of magnetic force and the beam. When the beam 114 is undeflected by a signal applied to terminal 120, the increasing magnetic field in the neighborhood of conductor 123 forces the electrons onto a path of progressively decreasing radius of curvature, as shown in FIG. 4, until they reach a point closest to conductor 123 after which they are deflected away from this conductor over a path symmetrical to the path of approach. If the beam has been originally skew to conductor 123, its magnetic deflection parallel to this conductor will be reduced and the electron path will be curved away from the median plane previously defined. A target electrode 115 of resistance material, provided with conductive terminal members 116', 116'' from which connections extend to an output circuit not further illustrated which may be similar to that of FIG. 1, is in the form of a strip concentric with conductor 123; strip 115 extends along a space curve in such manner as to intercept the electrons deflected by the magnetic field around conductor 123 in various positions of the beam.

From FIG. 3 it will be apparent that the point of impingement of the electrons upon target 115 shifts considerably with relatively small deviations of the beam position from the tube axis. This sensitivity, on the other hand, may result in a certain dispersion of the electrons at the target, as illustrated, especially in the case of beams of appreciable cross section. Since, however, the voltage appearing across terminals 116', 116'' will be determined by the means locus of impingement of the intercepted electrons, the output signal will not be materially affected by this dispersion.

A magnetic field similar to that existing around conductor 123 may also be produced by a stream of free electrons. This has been illustrated in FIG. 5 where, by way of further modification, two electron currents 223', 223'' parallel to a controlled beam 214 are generated by respective cathodes 227', 227'', anodes 228', 228'' and biasing batteries 229', 229''. Beam 214, after passing along or parallel to the axis of tube 210 through the field of a focusing coil 222, reaches a target electrode 215 after passing through a magnetic field which is zero along the tube axis and reaches maxima of opposite sign adjacent electron currents 223' and 223'', being thus similar to the field produced by magnets 23' and 23'' in FIG. 1. It may be mentioned that, as illustrated, the presence of beam 214 will also tend to deflect the electron currents 223', 223'' from their path parallel to the tube axis; such deflection may, however, be minimized by imparting a sufficiently high velocity to the electrons emitted by cathodes 227', 227''. This velocity may be controlled, for the purpose of varying the degree of deflection of beam 214, by means of potentiometers 230', 230''.

The invention is, of course, not limited to the specific embodiments described and illustrated but may be realized in various modifications, substitutions (e.g. of an electron beam similar to beam 223' for conductor 123), adaptations or combinations without departing from the spirit and scope of the appended claims.

I claim:

1. An electronic signal delay device comprising an elongated envelope, a source of electrons in said envelope, means for forming said electrons into a beam, a source of high-frequency signals, means controlled by said source of signals for deflecting said beam substantially parallel to a normal beam position, collector means for said elec-

4

trons in said envelope spaced from said deflecting means by a distance such that the transit time of said electrons therebetween equals at least a substantial fraction of an oscillatory cycle of said signals, and magnetic focusing means positioned between said deflecting means and said collector means for minimizing deviations of said beam from parallelism with said normal beam position.

2. A device according to claim 1 wherein said controlled means comprises source means for producing a steady transverse magnetic field, said field varying progressively in intensity and changing sign substantially in the region of the undeflected beam, and signal-response beam-deflecting means positioned ahead of said field.

3. A device according to claim 2 wherein said source means comprises a pair of inversely polarized magnets positioned on opposite sides of said region.

4. A device according to claim 1, further comprising signal-independent beam-deflecting means positioned between said focusing means and said collector means for causing divergence of the deflected beam from said normal beam position.

5. A device according to claim 4 wherein said beam-deflecting means comprises source means for producing a steady magnetic field ahead of said collector means.

6. A device according to claim 5 wherein said source means comprises a pair of inversely polarized magnets positioned on opposite sides of the region of the undeflected beam.

7. A device according to claim 5 wherein said source means comprises means for producing a current flow in a direction perpendicular to the path of the undeflected beam.

8. A device according to claim 5 wherein said controlled means is adapted to deflect said beam in one plane, said source means comprising means for producing current flow in said plane parallel to said normal beam position on opposite sides of the latter.

9. An electronic signal-delay device comprising an elongated envelope, a source of electrons in said envelope, means for forming said electrons into a beam, a source of high-frequency signals, signal-responsive beam-control means adjacent said source for deflecting said beam substantially parallel to itself and to a normal beam position, signal-independent beam-control means remote from said source for causing divergence of the deflected beam from said normal beam position, and collector means for said electrons in said envelope at a location beyond said signal-independent beam-control means.

10. A device according to claim 9 wherein each of said beam-control means comprises means for producing a steady magnetic field transverse to said normal beam position, said fields being parallel to each other and increasing progressively with distance from said normal beam position.

References Cited in the file of this patent

UNITED STATES PATENTS

2,263,376	Blumlein et al.	Nov. 18, 1941
2,361,998	Fleming-Williams	Nov. 7, 1944
2,455,977	Bocciarelli	Dec. 14, 1948
2,498,354	Bocciarelli	Feb. 21, 1950
2,824,997	Haeff	Feb. 25, 1958

FOREIGN PATENTS

| 429,916 | Great Britain | June 11, 1935 |

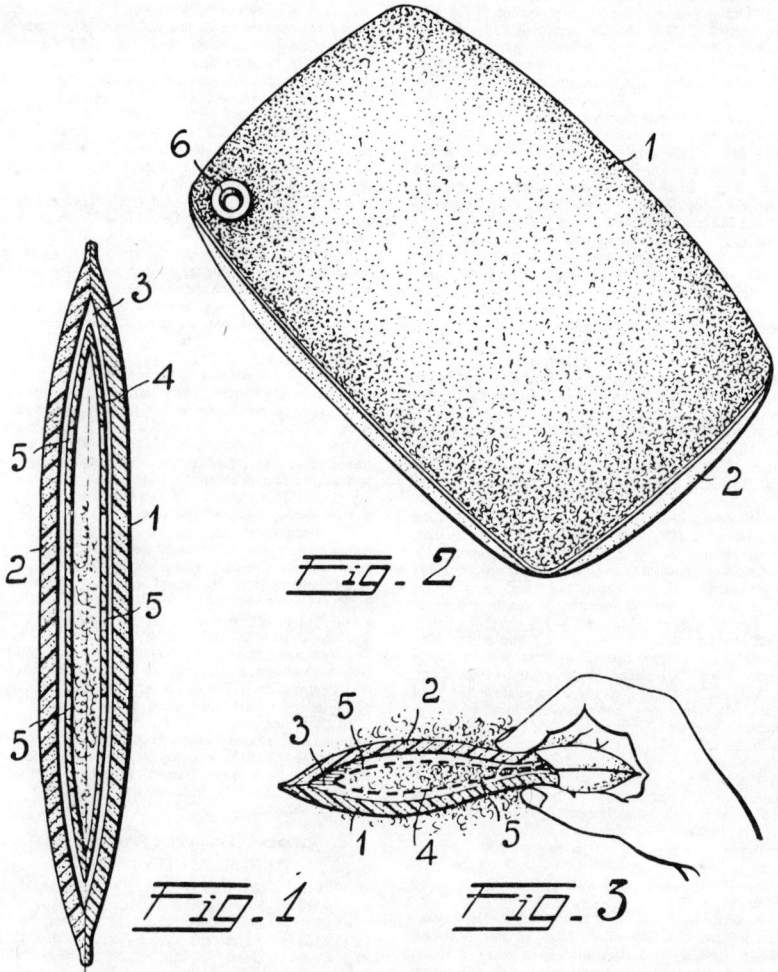

Fig. 2

Fig. 1

Fig. 3

INVENTOR:
Ada Bottino

BY

Karl F. Ross

AGENT.

1

2,899,780

SCOURING PAD CONTAINING DISPENSABLE SUBSTANCE

Ada Bottino, Rome, Italy

Application December 27, 1956, Serial No. 630,967

3 Claims. (Cl. 51—185)

This invention relates to an improvement in scouring pads, of the kind which is suitable for use in household operations, as for instance as a kitchen pad for washing dishes, walls and the like, as well as in industrial or similar operations.

Porous or permeable materials have been already used heretofore in the manufacture of scouring pads. It is well known in fact that porous natural or synthetic rubbers have found one or their main fields of application in this direction, and very satisfactory results in this field have been reached with the continuous improvements of plastic materials having a porous, cellular structure. It has also been proposed heretofore to combine such pads with a certain quantity of detergents, as for instance powdered soap, the pad being to this end designed in the form of a bag wherein the detergent is located. It is however known that such bag-shaped pads are affected by the drawback that, since during use the inner compartment thereof is flooded by an uncontrollable quantity of water causing the whole detergent to be brought into solution, the consumption of this latter occurs very quickly, whereas the formation of the washing foam is proceeding in a very irrational manner because of the disproportionate quantities of water and detergent.

It is an object of the invention to provide an improved pad of this class, in which the consumption of the added dispensible substance, such as detergents or the like, is positively reduced.

It is another object of the invention to provide a bag-shaped pad of the class described wherein the proportions of water and added substances in the inner of the bag may be regulated at will, so as to obtain the optimum solution for forming a rich foam. More particularly, it is an object of the invention to provide a pad in which the added substance may be admixed with or dispensed into the wash water, which has entered the inner compartment of the bag, in measured quantities.

According to an important feature of the invention, the pad comprises an outer bag element at least partly made of permeable material having a cellular structure, and an inner flexible dispenser of impermeable material containing the added substance, passages being provided in the walls of said dispenser in order to allow the added substance to be ejected upon a squeezing of the inner dispenser.

These and other objects and features of the invention will be better understood from the following detailed description, reference being taken to the accompanying drawing in which:

Fig. 1 is a cross-sectional view of a scouring pad according to the invention.

Fig. 2 is a perspective view of the pad; and

Fig. 3 is a partly cross-sectional and partly perspective view of the pad during the use, when squeezed by the user.

Referring now more particularly to the drawing, the bag-shaped pad is formed of two parts 1 and 2, made

2

of permeable plastic material having a cellular porous structure, which are connected along their peripheral rims by means of adhesive, electronic welding, stitching or the like. Thus an inner compartment 3 is formed which may be entered by an external liquid through the permeable walls of this compartment. A portion of the perimetral rim of the bag may be left open to facilitate the introduction into said compartment of an inner dispenser 4, which is formed of a flexible envelope, for instance of an inherently water impenetratable plastic material such as polyamide or polyvinyl plastic foils. This dispenser has a size which is slightly inferior to the overall size of the compartment 3 and is filled with a soluble cleansing agent, such as powdered soap. A series of holes 5 of limited size are provided in the walls of the dispenser.

An eyelet 6 is secured in a conventional manner to the pad.

It has been found to be very useful to make one of the parts 1 and 2, e.g. part 1, of a porous material having larger cells in its structure than the cellular material of part 2. Such large-cell materials are remarkably rougher to the touch than fine microcellular materials, so that one face of the pad may be used against surfaces needing a more intensive abrasive action, whereas the other face may be used for rubbing or polishing finer surfaces. It has been also found to be convenient to incorporate in part 1, during its manufacture, an abrasive powder substance, so as to increase the abrasive characteristics of this part.

As shown in Fig. 3. when the pad is squeezed with the hand a certain limited quantity of detergent is ejected from the dispenser into the compartment 3 through the holes 5, the compartment being flooded by water which penetrates into this compartment through the permeable parts 1 and 2 as soon as the pad is plunged into water or sprayed with water. The squeezing is repeated as long as necessary to obtain optimum foam.

The pad is thus impregnated with a foaming solution, yet the small size of the holes 5 in the dispensing envelope 4 as well as the impenetrability of the wall portions between these holes have the effect of keeping the interior of said dispensing envelope insulated from the surrounding water.

The added substance contained in envelope 4 may be an antiseptic or insecticide agent or a wax, and in the latter case of the side parts 1 and 2, e.g. part 1, may be manufactured with a fine cellular structure so as to be capable of uniformly distributing the wax even on delicate surfaces, as for instance on pieces of furniture, without injuring the same. At least one of parts 1 and 2 may be made of wholly impermeable material.

Instead of being incorporated in the structure of parts 1 or 2, the abrasive or emery materials may be adhered to the surface thereof with the aid of a solvent or a glue, or may be sprayed thereon in the course of the setting of the thermoplastic materials forming these parts 1 or 2.

Although it is preferable to have the rims of the pad completely closed after the introduction of the dispenser, at least the portion designed for the introduction of said dispenser 4 may be only temporarily closed, for instance by means of buttons, lacing or the like, in a conventional manner, so as to allow the dispenser to be replaced.

I claim:

1. A scouring pad comprising a bag consisting at least partly of permeable flexible material having a cellular structure, a flexible dispenser inside said bag, and a dispensable substance in said dispenser; said dispenser being made of an inherently water-impenetrable material provided with a plurality of permanently open apertures distributed around its periphery, said apertures

2,899,780

3

enabling the ejection of said substance upon a squeezing of said bag, said dispenser occupying a volume substantially smaller than the interior of said bag, said bag thus forming a water-penetrable compartment about at least a portion of said dispenser.

2. A scouring pad according to claim 1, wherein said bag consists of two substantially coextensive sheet members of permeable flexible material, said sheet members being joined together over the major part of their perimeter while forming a slit for the introduction of said dispenser.

3. A scouring pad according to claim 2, wherein said bag is provided with temporary closure means at said slit.

4

References Cited in the file of this patent

UNITED STATES PATENTS

574,449	Pritschau	Jan. 5,	1897
1,433,038	Rex	Oct. 24,	1922
1,854,415	Miller	Apr. 19,	1932
2,107,636	Kingman	Feb. 8,	1938
2,209,914	Gerber et al.	July 30,	1940
2,288,002	Kingman	June 30,	1942
2,375,585	Rimer	May 8,	1945
2,376,783	Kingman	May 22,	1945
2,650,158	Eastman	Aug. 25,	1953

United States Patent [19]

Boldt et al.

[11] **Des. 245,841**

[45] ** **Sept. 20, 1977**

[54] **TELEVISION RECEIVER**

[75] Inventors: **Melvin H. Boldt,** Glenview; **Richard K. Althans,** Long Grove, both of Ill.

[73] Assignee: **Zenith Radio Corporation,** Glenview, Ill.

[**] Term: **7 Years**

[21] Appl. No.: **666,711**

[22] Filed: **Mar. 15, 1976**

[51] Int. Cl. .... **D14—03**
[52] U.S. Cl. **D14/80**
[58] **Field of Search** D14/1, 2, 5–7, D14/14, 16, 17, 19–21, 23–25, 31–37, 39, 40, 68–77, 79–82, 84, 99

[56] **References Cited**

U.S. PATENT DOCUMENTS

D. 217,572	5/1970	James et al.	D14/80
D. 229,971	1/1974	Ito	D14/80
D. 239,325	3/1976	Boldt	D14/80

OTHER PUBLICATIONS

W. Bell & Co., 1975 Internat'l. Gift Collection Cat.,© 1974, p. 248, Item "I", #635 29K 9488.
Best Products, cat. recv'd, Apr. 23, 1975, p. 30, Item "A" RCA 12" black, white portable TV, #432652ECR6997.

Primary Examiner—Joel Stearman
Assistant Examiner—J. Corrigan
Attorney, Agent, or Firm—John J. Pederson

[57] **CLAIM**

The ornamental design for a television receiver, substantially as shown and described.

DESCRIPTION

FIG. 1 is a front, top and left side perspective view of a television receiver showing our new design;
FIG. 2 is a right side elevational view;
FIG. 3 is a top plan view; and
FIG. 4 is a rear elevational view.

Fig.1.

Fig.2.

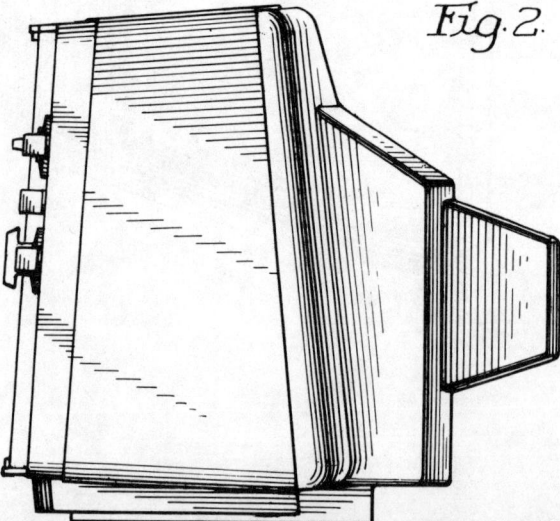

Fig. 3.

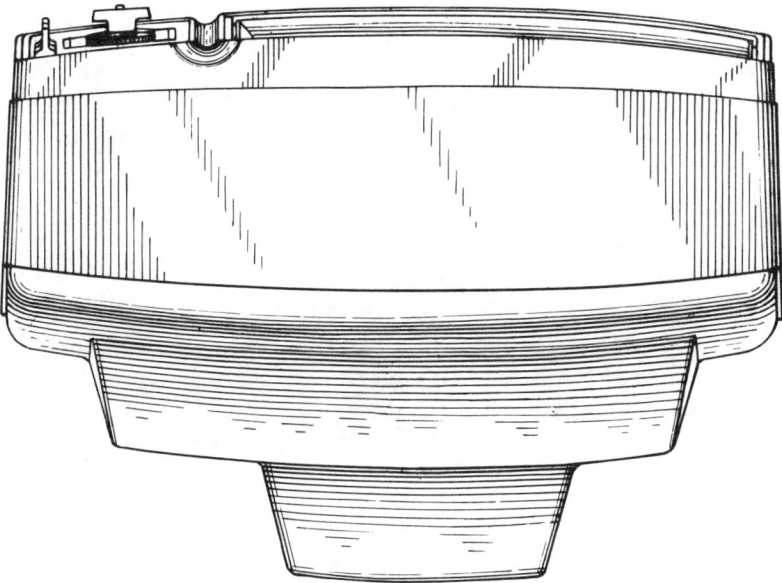

Fig. 4.

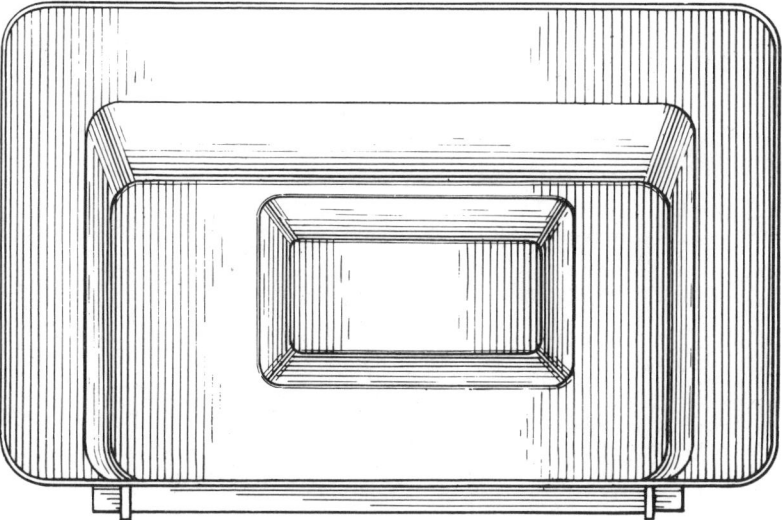

Standards for Drawings

Reproduced on the following pages is the entire *Guide for Patent Draftsmen,* available for $.65 from the Superintendent of Documents, U.S. Government Printing Office, Washington, D.C. 20402. It lists the regulations that apply to the standards for drawings submitted to the U.S. Patent and Trademark Office as part of a patent application. It also gives examples of how various elements of the invention should be rendered in the drawings.

If your drawings fail to conform to some of these standards, your application will still be accepted and examined, but the PTO will require new or corrected drawings before issuing a patent.

Selected Rules of Practice relating to Patent Drawings

THE DRAWINGS

35 U. S. C. 113. Drawings. *When the nature of the case admits, the applicant shall furnish a drawing.*

81. DRAWINGS REQUIRED. The applicant for patent is required by statute to furnish a drawing of his invention whenever the nature of the case admits of it; this drawing must be filed with the application. Illustrations facilitating an understanding of the invention (for example flow sheets in cases of processes, and diagrammatic views) may also be furnished in the same manner as drawings, and may be required by the Office when considered necessary or desirable.

No names or other identification will be permitted within the "sight" of the drawing, and applicants are expected to use the space above and between the hole locations to identify each sheet of drawings. This identification may consist of the attorney's name and docket number or the inventor's name and case number and may include the sheet number and the total number of sheets filed (for example, "sheet 2 of 4").

83. CONTENT OF DRAWING. (a) The drawing must show every feature of the invention specified in the claims. However, conventional features disclosed in the description and claims, where their detailed illustration is not essential for a proper understanding of the invention, should be illustrated in the drawing in the form of a graphical drawing symbol or a labeled representation (e.g. a labeled rectangular box).

(b) When the invention consists of an improvement on an old machine the drawing must when possible exhibit, in one or more views, the improved portion itself, disconnected from the old structure, and also in another view, so much only of the old structure as will suffice to show the connection of the invention therewith.

84. STANDARDS FOR DRAWINGS.

(a) **Paper and ink.** Drawings must be made upon pure white paper of a thickness corresponding to two-ply or three-ply bristolboard. The surface of the paper must be calendered and smooth and of a quality which will permit erasure and correction with India ink. India ink, or its equivalent in quality, must be used for pen drawings to secure perfectly black solid lines. The use of white pigment to cover lines is not acceptable.

(b) **Size of sheet and margins.** The size of a sheet on which a drawing is made must be exactly 8½ by 14 inches (21.6 by 35.6 cm.) One of the shorter sides of the sheet is regarded as its top. The drawing must include a top margin of 2 inches (5.1 cm) and bottom and side margins of one-quarter inch (6.4 mm) from the edges, thereby leaving a "sight" precisely 8 to 11¾ inches (20.3 by 29.8 cm.) Margin border lines are not permitted. All work must be included within the "sight." The sheets may be provided with two ¼-inch-diameter (6.4 mm.) holes having their centerlines spaced eleven-sixteenths inch (17.5 mm.) below the top edge and 2¾ inches (7.0 cm.) apart, said holes being equally spaced from the respective side edges.

(c) **Character of lines.** All drawings must be made with drafting instruments or by a process which will give them satisfactory reproduction characteristics. Every line and letter must be absolutely black and permanent; the weight of all lines and letters must be heavy enough to permit adequate reproduction. This direction applies to all lines however fine, to shading, and to lines representing cut surfaces in sectional views. All lines must be clean, sharp, and solid, and fine or crowded lines should be avoided. Solid black should not be used for sectional or surface shading. Freehand work should be avoided wherever it is possible to do so.

(d) **Hatching and shading.** (1) Hatching should be made by oblique parallel lines, which may be not less than about one-twentieth inch (1.3 mm) apart.

(2) Heavy lines on the shade side of objects should be used except where they tend to thicken the work and obscure reference characters. The light should come from the upper

left-hand corner at an angle of 45°. Surface delineations should be shown by proper shading, which should be open.

(e) Scale. The scale to which a drawing is made ought to be large enough to show the mechanism without crowding when the drawing is reduced in reproduction, and views of portions of the mechanism on a larger scale should be used when necessary to show details clearly; two or more sheets should be used if one does not give sufficient room to accomplish this end, but the number of sheets should not be more than is necessary.

(f) Reference characters. The different views should be consecutively numbered figures. Reference numerals (and letters, but numerals are preferred) must be plain, legible and carefully formed, and not be encircled. They should, if possible, measure at least one-eighth of an inch (3.2 mm) in height so that they may bear reduction to one twenty-fourth of an inch (1.1 mm); and they may be slightly larger when there is sufficient room. They must not be so placed in the close and complex parts of the drawing as to interfere with a thorough comprehension of the same, and therefore should rarely cross or mingle with the lines. When necessarily grouped around a certain part, they should be placed at a little distance, at the closest point where there is available space, and connected by lines with the parts to which they refer. They should not be placed upon hatched or shaded surfaces but when necessary, a blank space may be left in the hatching or shading where the character occurs so that it shall appear perfectly distinct and separate from the work. The same part of an invention appearing in more than one view of the drawing must always be designated by the same character, and the same character must never be used to designate different parts.

(g) Symbols, legends. Graphical drawing symbols and other labeled representations may be used for conventional elements when appropriate, subject to approval by the Office. The elements for which· such symbols and labeled representations are used must be adequately identified in the specification. While descriptive matter on drawings is not permitted, suitable legends may be used, or may be required in proper cases, as in diagrammatic views and flow sheets or to show materials or where labeled representations are employed to illustrate conventional elements. Arrows may be required, in proper cases, to show direction of movement. The lettering should be as large as, or larger than, the reference characters.

(i) Views. The drawing must contain as many figures as may be necessary to show the invention; the figures should be consecutively numbered if possible, in the order in which they appear. The figures may be plan, elevation, section, or perspective views, and detail views of portions or elements, on a larger scale if necessary, may also be used. Exploded views, with the separated parts of the same figure embraced by a bracket, to show the relationship or order of assembly of various parts are permissible. When necessary a view of a large machine or device in its entirety may be broken and extended over several sheets if there is no loss in facility of understanding the view (the different parts should be identified by the same figure number but followed by the letters, **a, b, c,** etc., for each part). The plane upon which a sectional view is taken should be indicated on the general view by a broken line, the ends of which should be designated by numerals corresponding to the figure number of the sectional view and have arrows applied to indicate the direction in which the view is taken. A moved position may be shown by a broken line superimposed upon a suitable figure if this can be done without crowding, otherwise a separate figure must be used for this purpose. Modified forms of construction can only be shown in separate figures. Views should not be connected by projection lines nor should center lines be used.

(j) Arrangement of views. All views on the same sheet must stand in the same direction and should if possible, stand so that they can be read with the sheet held in an upright position. If views longer than the width of the sheet are necessary for the clearest illustration of the invention, the sheet may be turned on its side so that the two-inch (5.1 cm) margin is on the righthand side. One figure must not be placed upon another or within the outline of another.

(k) Figure for Official Gazette. The drawing should, as far as possible, be so planned that one of the views will be suitable for publication in the Official Gazette as the illustration of the invention.

(l) Extraneous matter. An inventor's, agent's, or attorney's name, signature, stamp, or address, or another extraneous matter, will not be permitted upon the face of a drawing, within or without the margin, except that identifying indicia (attorney's docket number, inventor's name, number of sheets, etc.) should be placed within three-fourths, inch (19.1 mm) of the top edge and between the hole locations defined in paragraph (b) of this rule. Authorized security markings may be placed on the drawings provided they be outside the illustrations and are removed when the material is declassified.

(m) Transmission of drawings. Drawings transmitted to the Office should be sent flat, protected by a sheet of heavy binder's board, or may be rolled for transmission in a suitable mailing tube; but must never be folded. If received creased or mutilated, new drawings will be required. (See rule 152 for design drawings, 165 for plant drawings, and 174 for reissue drawings)

85. INFORMAL DRAWINGS. The requirements of rule 84 relating to drawings will be strictly enforced. A drawing not executed in conformity thereto, if suitable for reproduction, may be admitted, but in such case the drawing must be corrected or a new one furnished, as required. The necessary corrections or mounting will be made by the Office upon applicant's request or permission and at his expense. (See rules 21 and 165)

86. DRAFTSMAN TO MAKE DRAWINGS.

(a) Applicants are advised to employ competent draftsmen to make their drawings.

(b) The Office may furnish the drawings at the applicant's expense as promptly as its draftsmen can make them, for applicants who cannot otherwise conveniently procure them. (See rule 21)

88. USE OF OLD DRAWINGS. If the drawings of a new application are to be identical with the drawings of a previous application of the applicant on file in the Office, or with part of such drawings, the old drawings or any sheets thereof may be used if the prior application is, or is about to be, abandoned, or if the sheets to be used are cancelled in the prior application. The new application must be accompanied by a letter requesting the transfer of the drawings, which should be completely identified.

123. Amendments to the drawing.

(a) No change in the drawing may be made except by permission of the Office. Permissible changes in the construction shown in any drawing may be made only by the Office. A sketch in permanent ink showing proposed changes, to become part of the record, must be filed. The paper requesting amendments to the drawing should be separate from other papers.

(b) Substitute drawings will not ordinarily be admitted in any case unless required by the Office.

DESIGN PATENTS

152. DRAWING. The design must be represented by a drawing made in conformity with the rules laid down for drawings of mechanical inventions and must contain a sufficient number of views to constitute a complete disclosure of the appearance of the article. Appropriate surface shading must be used to show the character or contour of the surfaces represented.

HEAVY
LINES

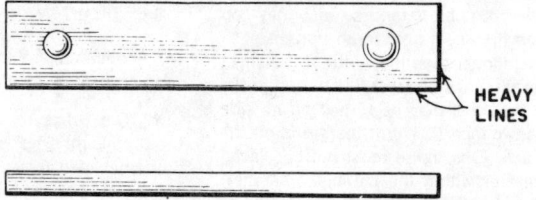

ALWAYS USE PLAIN BLOCK LETTERING
FOR LEGENDS NAMES, ETC.

SOME STYLES OF LETTERING
USED ON PATENT DRAWINGS

WATER
INSULATION   1234567890
COPPER
OIL

Fig. 1.

FIG. 1.

Fig. 1.

Fig. 1

ALL FIGS. MUST BE SEPARATELY NUMBERED

THE LIGHT COMES
FROM THE UPPER
LEFT-HAND CORNER
AT AN ANGLE
OF 45°

ALWAYS MAKE
SHADE LINES
ON SHADOW SIDE

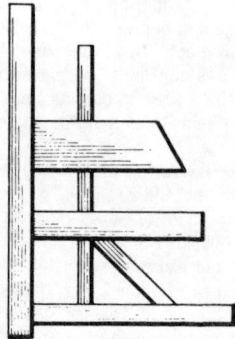

Letters and figures of reference must be carefully formed. Several types of lettering and figure marks are shown, however, the draftsman may use any style of lettering that he may choose.

Place heavy lines on the shade side of objects, assuming that the light is coming from the upper left-hand corner at an angle of 45°. Make these heavy lines the same weight throughout the various views on the drawing.

Descriptive matter is not permitted on patent drawings. Legends may be applied when necessary but only plain black lettering should be used.

The different views should be consecutively numbered.

SURFACE SHADING FOR
VARIOUS SIZES OF PIPES & SHAFTS

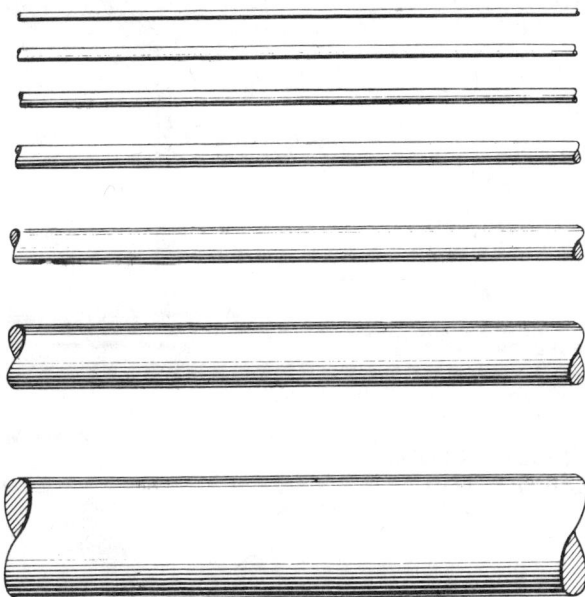

SURFACE SHADING FOR SPHERICAL OBJECTS

Surface delineations should be shown by proper shading. The figures show various types of surface shading. The amount of shading necessary depends on the size of the diameter of the shaft, etc. Note that a single heavy line on the shadow side is sufficient shading for small pipes, rods and shafts. When more than one shade line is used on cylindrical surfaces, the shading is blended from the second line. Note that the outer line is a light line. This rule on shading applies to spherical as well as cylindrical surfaces.

Make all lines clear and sharp so that they will reproduce properly.

India ink, or its equivalent in quality, must be used for pen drawings to secure perfectly black solid lines.

**SHADING FOR A BLOCK
IN PERSPECTIVE**

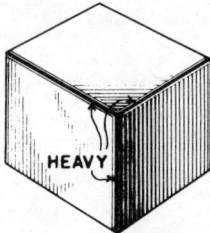

**SURFACE SHADING
ILLUSTRATING A MIRROR**

HEAVY

**NOTE—THE HEAVY SHADE LINES
ARE PLACED ON THE EDGES
CLOSEST TO THE EYE**

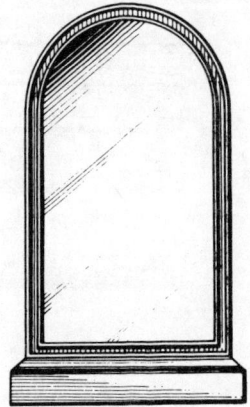

**RECTANGULAR BLOCK
IN PERSPECTIVE**

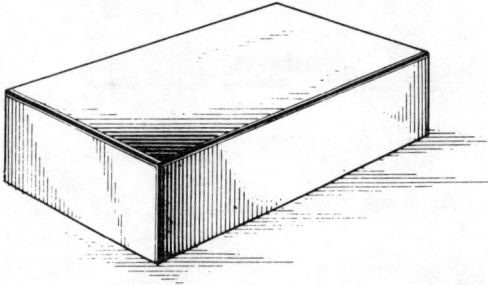

ROUND MIRROR

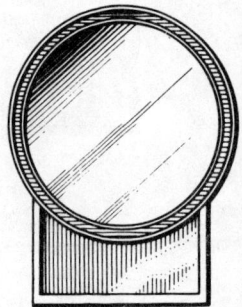

Heavy shade lines on perspective views are placed on the edges closest to the eye. The rule of the light coming from the upper left-hand corner at a 45° angle does not apply to perspective views. If a very light line is placed either side of the heavy line, a more finished appearance of the article is obtained. The addition of horizontal ground line as shown in the lower perspective view of the block emphasizes support for the same.

The appearance of a mirror or shiny surfaces can be illustrated by the oblique shading shown on the two views on the right-hand side of the page.

NUMERALS MUST BE PLACED AS CLOSE AS POSSIBLE TO THE PART TO WHICH THEY REFER

24   22
25   23
26
27   28

NEEDLE VALVE

WOOD SCREW

SHADING FOR ROUND HANDLES, ETC.

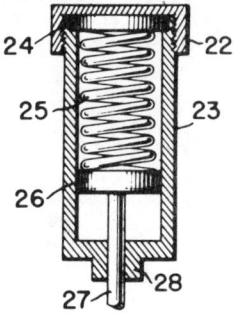

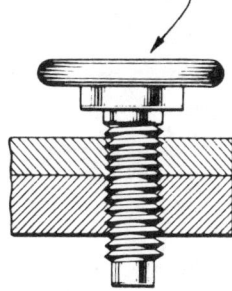

CYLINDRICAL SHADING CONVENTIONAL

CYLINDRICAL SHADING HIGH LIGHT

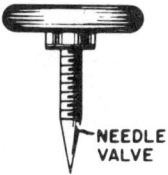

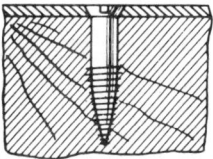

Reference characters should be placed at a little distance from the parts to which they refer. They should be connected with these parts by a short lead line, never by a long lead line. When necessary blank spaces must be left on shaded and hatched areas for applying the numerals.

Use wood graining sparingly on parts of wood in section. Excessive wood graining is objectionable as it blurs the view and is very confusing.

Various methods of shading are shown, however, the conventional surface shading should be used until the draftsman has obtained enough experience to attempt the more involved types of shading.

SURFACE SHADING ON BEVEL EDGES

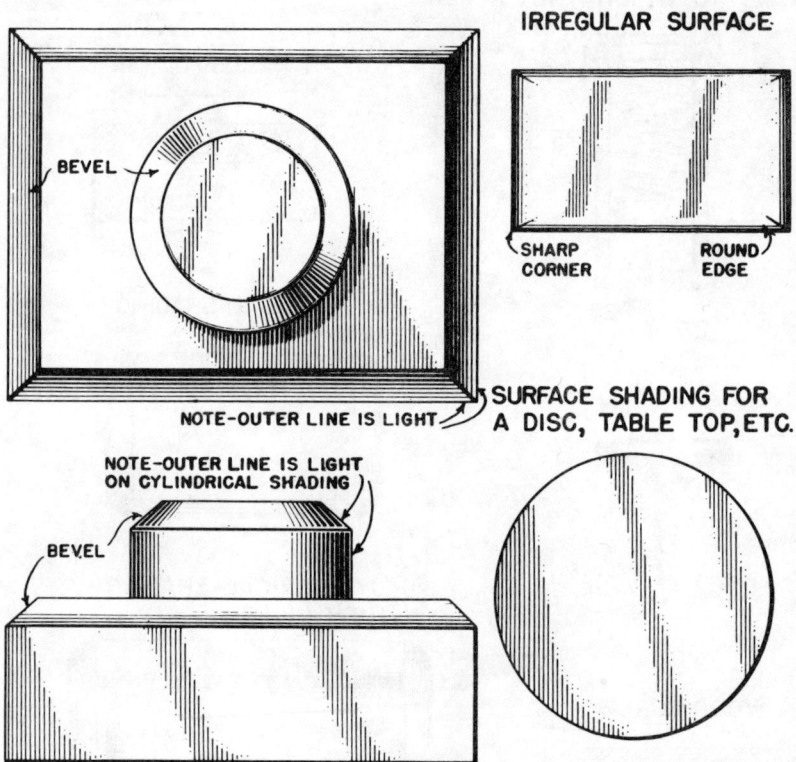

BEVEL

IRREGULAR SURFACE

SHARP CORNER   ROUND EDGE

NOTE-OUTER LINE IS LIGHT

SURFACE SHADING FOR A DISC, TABLE TOP, ETC.

NOTE-OUTER LINE IS LIGHT ON CYLINDRICAL SHADING

BEVEL

Inclined surfaces are distinguished from flat surfaces by using the shading shown on the illustration in the upper left-hand corner of the page. You will note that the outer line is always a light line. This gives a slanting effect to the surface as the heavy line is placed on the edge of the upper plane. The surface shading is blended from this heavy line giving the desired appearance.

The other figures on the page show various methods of shading. Flat, shiny surfaces may be shown as illustrated in the circular figure.

The scale to which a drawing is made ought to be large enough to show the mechanism without crowding.

THREADS-CONVENTIONAL METHOD

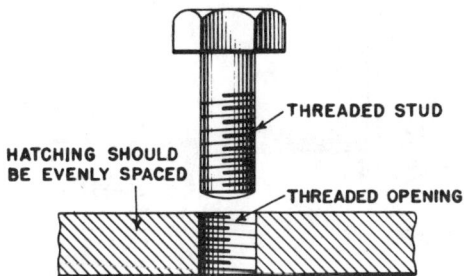

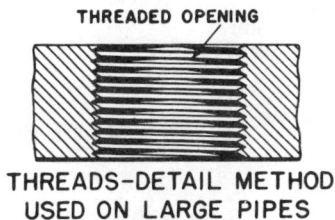

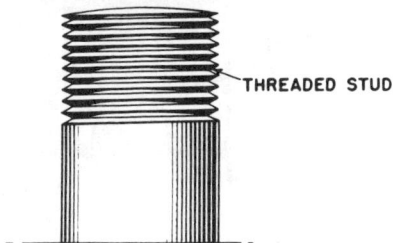

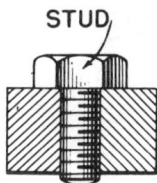

BALL

THREADED STUD

HATCHING SHOULD
BE EVENLY SPACED

THREADED OPENING

THREADED OPENING

SOCKET

THREADS-DETAIL METHOD
USED ON LARGE PIPES

THREADED STUD

STUD

Several methods of illustrating threads are shown in the figures above. The conventional thread may be shown on small bolts and openings. The detail method should be used on large pipes and threaded portions. Solid black shading as shown is very effective in illustrating the threads but care should be used in applying same.

Convex and concave surfaces are defined by the shading shown in the illustrations of the ball and socket.

Plan the views properly so that one figure is not placed upon another or within the outline of another.

BEVEL GEARS

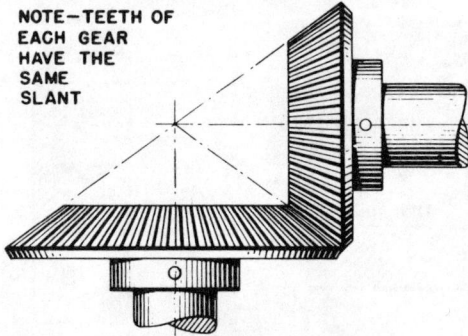

NOTE—TEETH OF
EACH GEAR
HAVE THE
SAME
SLANT

NOTE—ALL TEETH CONVERGE IN A CENTRAL
POINT.—BROKEN LINES ARE FOR INSTRUCTION
PURPOSES AND ARE NOT TO BE PLACED ON DRAWINGS

BALL BEARING

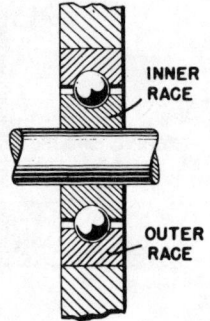

INNER
RACE

OUTER
RACE

TOP PLAN VIEW

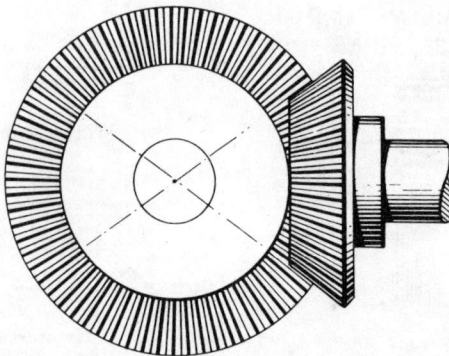

ROLLER BEARING

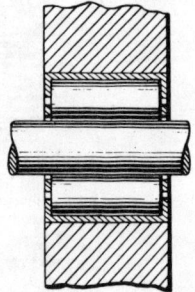

The conventional method of illustrating bevel gears is clearly shown on the two figures on the left-hand side of the page. Particular care must be given to the correct spacing between the gear teeth and also to the weight of the shade lines used. Both must be correctly shown to obtain the desired effect.

Two types of bearings are also shown. The roller bearing is clearly disclosed by the use of the conventional cylindrical shading. The fanciful black shading shown on the ball bearing is very effective in bringing out the idea of an object being shiny as well as round.

The use of white pigment to cover lines is not acceptable.

SPUR GEAR   **HELICAL GEAR**

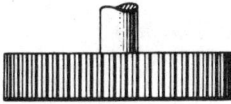
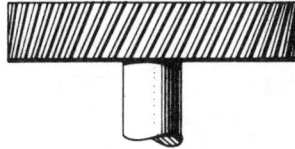

TWO SPUR GEARS IN MESH

WORM

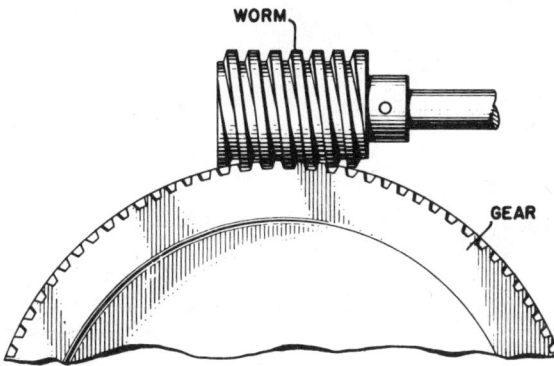

GEAR

The conventional method of showing a spur gear and a helical gear is shown on the two illustrations at the top of the page. The proper spacing between the gear teeth is essential in illustrating gears in mesh as shown on the central figure.

A worm and worm gear in mesh is clearly illustrated in the lower figure on the page. Do not add the legends on the drawing.

Every line and letter must be absolutely black. This direction applies to all lines however fine, to shading, and to lines representing cut surfaces in sectional views.

PERSPECTIVES

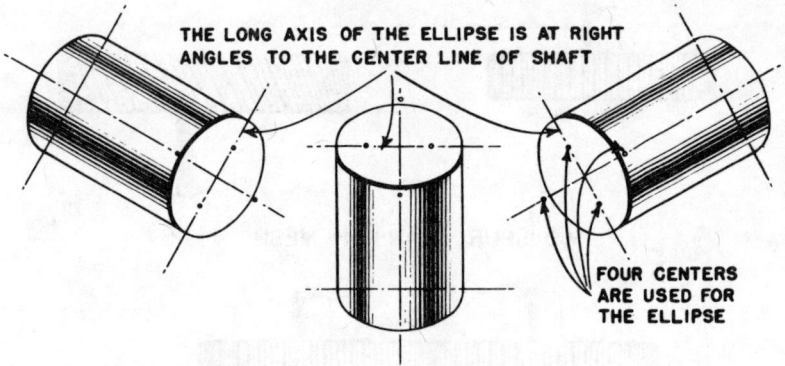

THE LONG AXIS OF THE ELLIPSE IS AT RIGHT
ANGLES TO THE CENTER LINE OF SHAFT

FOUR CENTERS
ARE USED FOR
THE ELLIPSE

THE LONG AXIS OF AN ELLIPSE
ON A HORIZONTAL SURFACE
IS ALWAYS HORIZONTAL

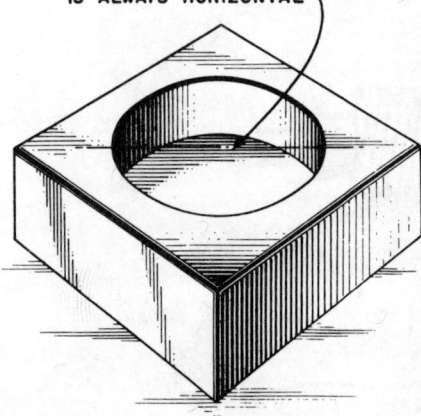

RADIO TUBE

GLASS SHOULD
BE SHOWN
WITH HIGH
LIGHT LINES
TO SHOW A
CONTRAST
WITH OTHER
MATERIAL

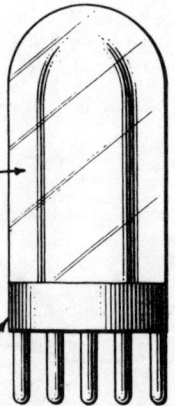

The four figures in perspective clearly explain the fundamental rules for determining the position of the long axis of the ellipse. Do not add the center lines as they are not permitted on patent drawings. These have been shown for instructive purposes only.

Different types of shading are used when it is desired to show a contrast between materials as shown in the illustration of the radio tube.

All drawings must be made with drafting instruments and every line and letter must be absolutely black. Free-hand work should be avoided wherever it is possible to do so.

LINK CHAIN
-SMALL-

LINK CHAIN
-LARGE DETAIL-

GRINDING WHEEL

TWO METHODS OF ILLUSTRATING FABRIC

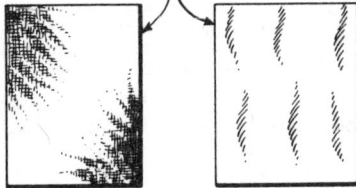

ABRASIVE MATERIAL MUST BE STIPPLED

LIGHT

ANOTHER METHOD OF ILLUSTRATING A CONICAL SURFACE

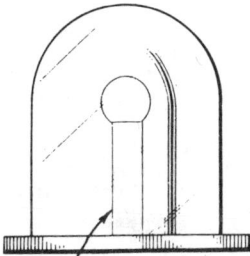

ELEMENTS BEHIND GLASS ARE SHOWN BY LIGHT LINES

Two illustrations of link chains are shown at the top of the page, the size of the view being the guiding factor in determining the correct showing.

Abrasive material must be stippled as shown in the illustration of the grinding wheel. Irregular surfaces and objects that are impossible to properly show up with line shading must be stippled to bring out the desired effect.

Free-hand shading should be used to designate fabric material.

All elements behind glass should be shown in light full lines. The light oblique shade lines across the glass give the desired effect.

SYMBOLS FOR DRAFTSMEN

Rule 84 (g) states that graphical symbols for conventional elements may be used on the drawing when appropriate, subject to approval by the Office. The symbols and other conventional devices which follow have been and are approved for such use. This collection does not purport to be exhaustive, other standard and commonly used symbols will also be acceptable provided they are clearly understood, are adequately identified in the specification as filed, and do not create confusion with other symbols used in patent drawings.

It should be noted that the American National Standards Institute Inc., 1430 Broadway, New York, N.Y. 10018, publishes a series of publications relating to graphic symbols under its Y32 and Z32 headings, the Office calls attention of patent applicants to these symbols for their consideration and use where appropriate in patent drawings. The listed publications have been reviewed by the Office and the symbols therein are considered to be generally acceptable in patent drawings. Although the Office will not "approve" all of the listed symbols as a group because their use and clarity must be decided on a case-by-case basis, these publications may be used as guides when selecting graphic symbols. Overly specific symbols should be avoided. Symbols with unclear meanings should be labeled for clarification. As noted in Rule 84 (g), the Office will retain final authority to approve the use of any particular symbols in any particular case.

The reviewed publications are as follows:

Y32.2—1970. Graphic Symbols for Electrical and Electronics Diagram $11.50
32.10—Graphic Symbols for Fluid Power Diagrams 3.00
Y32.11—1961. Graphic Symbols for Process Flow Diagrams in the Petroleum and Chemical Industries 2.00
Y32.14—1962. Graphic Symbols for Logic Diagrams 4.75
Z32.2.3—1949 (R1953). Graphic Symbols for Pipe Fittings, Valves and Piping . 2.00
Z32.2.4—1949 (R1953). Graphic Symbols for Heating, Ventilating and Air Conditioning 2.00
Z32.2.6—1950. Graphic Symbols for Heat-Power Apparatus 2.00

NOTES: In general, in lieu of a symbol, a conventional element, combination or circuit may be shown by an appropriately labeled rectangle, square, or circle; abbreviations should not be used unless their meaning is evident and not confusing with the abbreviations used in the suggested symbols. In the electrical symbols an arrow through an element indicates variability thereof, see for example symbols 2, 6, 12; dotted line connection of arrows indicates ganging thereof, see symbol 6; inherent property (as resistance) may be indicated by showing symbol (for resistor) in dotted lines.

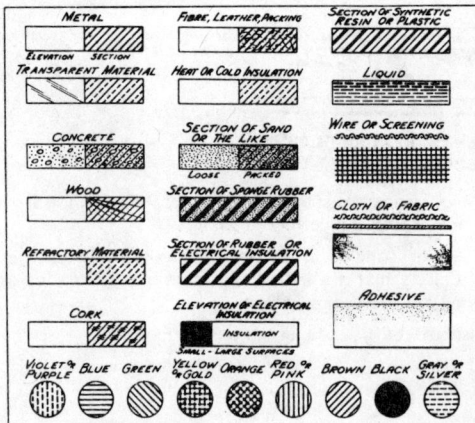

Electrical Symbols

RESISTOR	VARIABLE RESISTOR	POTENTIOMETER	RHEOSTATS	CONDENSERS	GANGED VARIABLE CONDENSERS
1	2	3	4	5	6
INDUCTORS	INDUCTOR ADJUSTABLE CORE	INDUCTOR OR REACTOR POWDERED MAGNETIC CORE	TRANSFORMER SATURABLE CORE	TRANSFORMER AIR CORE	VARIABLE TRANSFORMER
7	8	9	10	11	12
TRANSFORMER MAGNETIC CORE	AUTO-TRANSFORMER ADJUSTABLE	CROSSED AND JOINED WIRES	MAIN CIRCUITS / SHUNT OR CONTROL CIRCUITS	FUSE	COAXIAL CABLES
13	14	15	16	17	18
SHIELDING	BATTERY	THERMOELEMENT	BELL	AMMETER	MILLIAMMETER
19	20	21	22	23	24
VOLTMETER	GALVANOMETER	WATTMETER	SWITCH	DOUBLE POLE SWITCH	DOUBLE POLE DOUBLE THROW SWITCH
25	26	27	28	29	30
PUSH BUTTON TWO POINT MAKE	SELECTOR OR CONNECTOR OR FINDER SWITCH	CIRCUIT BREAKER OVERLOAD	RELAY	POLARIZED RELAY	DIFFERENTIAL RELAY
31	32	33	34	35	36
ANNUNCIATORS SIDE FRONT	DROP ANNUNCIATOR	DRUM TYPE SWITCH OR CONTROL	COMMUTATOR MOTOR OR GENERATOR	REPULSION MOTOR	INDUCTION MOTOR THREE PHASE SQUIRREL CAGE
37	38	39	40	41	42
INDUCTION MOTOR PHASE WOUND SECONDARY	SYNCHRONOUS MOTOR OR GEN. THREE PHASE	MOTOR GENERATOR	ROTARY CONVERTER THREE PHASE	FREQUENCY CHANGER THREE PHASE	TROLLEYS
43	44	45	46	47	48
THIRD RAIL SHOE	RECEIVERS	TRANSMITTER OR MICROPHONE	TELEPHONE HOOK	TELEGRAPH KEY	SWITCH BOARD PLUG AND JACK
49	50	51	52	53	54

Electrical Symbols — continued

PHONOGRAPH PICK UP 55	DYNAMIC SPEAKER 56	ANTENNA 57	LOOP ANTENNA 58	GROUND 59	SPARK GAP 60
LIGHTNING ARRESTER 61	DETECTOR or RECTIFIER — Anode — Cathode — GENERIC 62	DETECTOR or RECTIFIER — Anode — Cathode — CRYSTAL 63	PIEZOELECTRIC CRYSTAL 64	INCANDESCENT LAMP 65	MERCURY ARC RECTIFIER 66
ENVELOPE GAS FILLED 67	DIODE 68	TRIODE 69	PENTODE INDIRECTLY HEATED CATHODE 70	TRANSISTOR EMITTER COLLECTOR BASE 71	TRANSISTOR EMITTER COLLECTOR BASE 72
TRANSISTOR NPN JUNCTION TYPE 73	TRANSISTOR PNP JUNCTION TYPE 74	AMPLIFIER A 75	THERMIONIC FULL WAVE RECTIFIER 76	FULL WAVE RECTIFIER GAS FILLED 77	PHOTOELECTRIC CELL 78
GLOW DISCHARGE TUBE 79	X-RAY TUBE 80	CATHODE RAY TUBE 81	SPOT WELDING 82	DEPOSIT WELDING 83	

Mechanical Symbols

CONDUIT CROSSING AND INTERSECTING 1	SECTIONS LARGE ENDS ROD PIPE 2	SCREW THREAD 3	CLUTCH 4	FRICTION CLUTCH 5	BRAKE 6
FLEXIBLE COUPLING 7	FLUID COUPLING 8	SPROCKET and CHAIN 9		SPUR GEARS 10	BEVEL GEARS 11
WORM GEAR 12	SPUR GEARS SIDE VIEW 13	WELDS PLAN SECTION 14	SPOT WELD 15	INJECTOR NOZZLE 16	FIXED RESISTANCE 17

Mechanical Symbols — continued

Variable Resistance — 18	Pump — 19	Constant Delivery Pump — 20	Variable Delivery Pump — 21	Reversible Constant Delivery Pump — 22	Reversible Variable Delivery Pump — 23
Gear Pump — 24	Rotary Sliding Vane Pump — 25	Centrifugal Pump — 26	Lift Pump — 27	Force Pump — 28	Pneumatic Discharge Pump — 29
Air Lift Pump — 30	Ram — 31	Jet — 32	Steam Accumulator — 33	Mechanical Pressure Accumulator — 34	Air Pressure Accumulator — 35
Reservoir — 36	Motor — 37	Constant Speed Motor — 38	Variable Speed Motor — 39	Reciprocating Differential Motor — 40	Reciprocating Non-Differential Motor — 41
Gas Engine Two-Cycle — 42	Gas Engine Four-Cycle — 43	Diesel Engine Two-Cycle — 44	Diesel Engine Four Cycle — 45	Turbine — 46	Rocket Motor Fluid Fuel — 47
Rocket Motor Solid Fuel — 48	Jet Motor — 49	Turbo-Jet — 50	Boiler — 51	Fire Tube Boiler — 52	Flue Boiler — 53
Water Tube Boiler — 54	Jet Condenser — 55	Surface Condenser (Steam / Water) — 56	Jet Heater — 57	Surface Heater (Water / Steam) — 58	Valve — 59
Throttle Valve — 60	Check Valve — 61	Pressure Relief Valve — 62	Constant Pressure Outlet Valve — 63	Constant Pressure Inlet Valve — 64	Reducing Valve — 65
Three-Way Valve — 66	Distributing Valve — 67	Thermostatic Valve — 68	Bi-Metallic Thermostat — 69	Filter — 70	Heat Exchanger — 71

APPENDIX B

Patent Libraries

The libraries listed below have printed copies of U.S. patents arranged in numerical order:

Albany, New York	University of the State of New York
Atlanta, Georgia	Georgia Tech Library *
Birmingham, Alabama	Public Library
Boston, Massachusetts	Public Library
Buffalo, New York	Buffalo and Erie County Public Library
Chicago, Illinois	Public Library
Cincinnati, Ohio	Public Library
Cleveland, Ohio	Public Library
Columbus, Ohio	Ohio State Library
Dallas, Texas	Public Library
Denver, Colorado	Public Library
Houston, Texas	Fondren Library, Rice University
Kansas City, Missouri	Linda Hall Library *

* *Collection incomplete*

Los Angeles, California	Public Library
Madison, Wisconsin	State Historical Society of Wisconsin
Milwaukee, Wisconsin	Public Library
Newark, New Jersey	Public Library
New York, New York	Public Library
Philadelphia, Pennsylvania	Franklin Institute
Pittsburgh, Pennsylvania	Carnegie Hall
Providence, Rhode Island	Public Library
Raleigh, North Carolina	D. H. Hill Library, North Carolina State University
St. Louis, Missouri	Public Library
Seattle, Washington	University of Washington Engineering Library
Stillwater, Oklahoma	Oklahoma A&M College Library
Sunnyvale, California	Public Library **
Toledo, Ohio	Public Library

** *Arranged by subject matter, collection dates from January 2, 1962.*

APPENDIX C

U.S. Department of Commerce and Small Business Administration Field Offices

United States Department of Commerce Field Offices

The field offices listed below are arranged alphabetically according to city.

ALBUQUERQUE, NEW
MEXICO 87102
505 Marquette Ave., NW,
Suite 1015

ANCHORAGE, ALASKA 99501
412 Hill Bldg., 632 Sixth Ave.

ATLANTA, GEORGIA, 30309
Suite 600, 1365 Peachtree St.,
N.E.

BALTIMORE, MARYLAND
21202
415 U.S. Customhouse
Gay & Lombards Streets

BIRMINGHAM, ALABAMA
35205
Suite 200-201, 908 S. 20th St.

BOSTON, MASSACHUSETTS
02116
10th Floor
441 Stuart St.

BUFFALO, NEW YORK 14202
1312 Federal Building
111 W. Huron St.

CHARLESTON, WEST
VIRGINIA 25301
3000 New Federal Office
Building
500 Quarrier St.

CHEYENNE, WYOMING 82001
6022 O'Mahoney Federal
Center
2120 Capitol Ave.

CHICAGO, ILLINOIS 60603
1406 Mid Continental Plaza
Bldg.
55 E. Monroe St.

CINCINNATI, OHIO 45202
10504 Federal Office Bldg.
550 Main St.

CLEVELAND, OHIO 44114
Room 600
666 Euclid Avenue

COLUMBIA, SOUTH
CAROLINA 29204
2611 Forst Drive
Forest Center

DALLAS, TEXAS 75242
Room 7A5, 1100 Commerce St.

DENVER, COLORADO 80202
Room 165, New Customhouse
19th & Stout St.

DES MOINES, IOWA 50309
609 Federal Building
210 Walnut St.

DETROIT, MICHIGAN 48226
445 Federal Bldg.
231 W. Lafayette

GREENSBORO, N. CAROLINA
27402
203 Federal Bldg.
West Market St.
P.O. Box 1950

HARTFORD, CONNECTICUT
06103
Room 610-B, Federal Office
Bldg.
450 Main St.

HONOLULU, HAWAII 96813
286 Alexander Young Bldg.
1015 Bishop St.

HOUSTON, TEXAS 77002
2625 Federal Bldg., Courthouse
515 Rusk St.

KANSAS CITY, MISSOURI
64106
Room 1840, 601 E. 12th St.

LOS ANGELES, CALIFORNIA
90049
Room 800, 11777 San Vicente
Blvd.

MEMPHIS, TENNESSEE 38103
Room 710, 147 Jefferson Ave.

MIAMI, FLORIDA 33130
Room 821, City National
Bank Bldg.
25 W. Flagler St.

MILWAUKEE, WISCONSIN
53202
Federal Bldg./U.S. Courthouse
  517 E. Wisconsin Ave.

MINNEAPOLIS, MINNESOTA
55401
  218 Federal Bldg.
  110 S. 4th St.

NEWARK, NEW JERSEY 07102
  4th Floor, Gateway Bldg.
  Market St. & Penn Plaza

NEW ORLEANS, LOUISIANA
70130
  432 International Trade Mart
  No. 2 Canal St.

NEW YORK, NEW YORK 10007
  37th Floor, Federal Office Bldg.
  26 Federal Plaza, Foley Square

PHILADELPHIA,
PENNSYLVANIA 19106
  9448 Federal Bldg.
  600 Arch St.

PHOENIX, ARIZONA 85004
  508 Greater Arizona Savings
    Bldg.
  112 North Central Avenue

PITTSBURGH,
PENNSYLVANIA 15222
  2002 Federal Bldg.
  1000 Liberty Ave.

PORTLAND, OREGON 97204
  Room 618
  1220 S. W. 3rd Avenue

RENO, NEVADA 89502
  2028 Federal Bldg.
  300 Booth St.

RICHMOND, VIRGINIA 23240
  8010 Federal Bldg.
  400 N. 8th St.

ST. LOUIS, MISSOURI 63105
  120 S. Central Avenue

SALT LAKE CITY, UTAH 84138
  8010 Federal Bldg.
  125 S. State St.

SAN FRANCISCO,
CALIFORNIA 94102
  Federal Bldg., Box 36013
  450 Golden Gate Ave.

SAN JUAN, PUERTO RICO
00918
  Room 659 Federal Bldg.

SAVANNAH, GEORGIA 31402
  235 U.S. Courthouse &
  P.O. Bldg., 125-29 Bull St.

SEATTLE, WASHINGTON
98109
  Room 706, Lake Union Bldg.
  1700 Westlake Avenue North

Small Business Administration Field Offices

The field offices listed below are grouped by geographic regions.

Boston	Massachusetts 02114, 150 Causeway Street
Holyoke	Massachusetts 01040, 302 High Street
Augusta	Maine 04330, 40 Western Avenue, Room 512
Concord	New Hampshire 03301, 55 Pleasant Street
Hartford	Connecticut 06103, One Financial Plaza
Montpelier	Vermont 05602, 87 State Street, P.O. Box 605
Providence	Rhode Island 02903, 57 Eddy Street
New York	New York 10007, 26 Federal Plaza, Room 3214
Albany	New York 12207, Twin Towers Building, Room 922
Elmira	New York 14904, 180 State Street, Room 412
Hato Rey	Puerto Rico 00918, Federal Office Building, Carlos Chardon Avenue
Newark	New Jersey 07102, 970 Broad Street, Room 1635
Camden	New Jersey 08104, East Davis Street
Syracuse	New York 13202, 100 South Clinton Street, Room 1073
Buffalo	New York 14202, 111 West Huron Street
St. Thomas	Virgin Islands 00801, Franklin Building
Philadelphia	Bala Cynwyd, Pennsylvania 19004, One Bala Cynwyd Plaza
Harrisburg	Pennsylvania 17108, 1500 North Second Street
Wilkes-Barre	Pennsylvania 18702, 20 North Pennsylvania Avenue
Baltimore	Towson, Maryland 21204, 7800 York Road
Wilmington	Delaware 19801, 844 King Street
Clarksburg	West Virginia 26301, 109 N. 3rd Street
Charleston	West Virginia 25301, Charleston National Plaza, Suite 628
Pittsburgh	Pennsylvania 15222, 1000 Liberty Avenue
Richmond	Virginia 23240, 400 N. 8th Street, Room 3015
Washington	D.C. 20417, 1030 15th Street, NW., Suite 250

Atlanta	Georgia 30309, 1720 Peachtree Road, NW., Suite 600
Biloxi	Mississippi 39530, 111 Fred Haise Boulevard
Birmingham	Alabama 35205, 908 South 20th Street
Charlotte	North Carolina 28202, 230 South Tryon Street, Suite 700
Greenville	North Carolina 27834, 215 South Evans Street
Columbia	South Carolina 29201, 1801 Assembly Street
Coral Gables	Florida 33134, 2222 Ponce de Leon Boulevard
Jackson	Mississippi 39201, 200 East Pascagoula Street
Jacksonville	Florida 32202, 400 W. Bay Street
West Palm Beach	Florida 33402, 701 Clematis Street
Tampa	Florida 33607, 1802 North Trask Street, Suite 203
Louisville	Kentucky 40202, 600 Federal Place, Room 188
Nashville	Tennessee 37219, 404 James Robertson Parkway, Suite 1012
Knoxville	Tennessee 37902, 502 South Gay Street, Room 307
Memphis	Tennessee 38103, 167 North Main Street
Chicago	Illinois 60604, 219 South Dearborn Street
Springfield	Illinois 62701, 1 North Old State Capitol Plaza
Cleveland	Ohio 44199, 1240 East 9th Street, Room 317
Columbus	Ohio 43215, 34 North High Street
Cincinnati	Ohio 45202, 550 Main Street, Room 5524
Detroit	Michigan 48226, 477 Michigan Avenue
Marquette	Michigan 49885, 540 West Kaye Avenue
Indianapolis	Indiana 46204, 575 North Pennsylvania Street
Madison	Wisconsin 53703, 122 West Washington Avenue, Room 713
Milwaukee	Wisconsin 53233, 735 West Wisconsin Avenue
Eau Claire	Wisconsin 54701, 500 South Barstow Street, Room B9AA
Minneapolis	Minnesota 55402, 12 South Sixth Street
Dallas	Texas 75202, 1100 Commerce Street
Albuquerque	New Mexico 87110, 5000 Marble Avenue, NE.
Houston	Texas 77002, 1 Allen Center, Suite 705
Little Rock	Arkansas 72201, 611 Gaines Street, P.O. Box 1401
Lubbock	Texas 79401, 1205 Texas Avenue
El Paso	Texas 79902, 4100 Rio Bravo, Suite 300

Lower Rio Grande Valley	Harlingen, Texas 78550, 222 East Van Buren, Suite 500
Corpus Christi	Texas 78408, 3105 Leopard Street, P.O. Box 9253
Marshall	Texas 75670, 100 South Washington Street, Room G12
New Orleans	Louisiana 70113, 1001 Howard Avenue
Shreveport	Louisiana 71163, 500 Fannin Street
Oklahoma City	Oklahoma 73102, 200 NW. 5th Street
San Antonio	Texas 78206, 727 East Durango, Room A–513
Kansas City	Missouri 64106, 1150 Grand Avenue
Des Moines	Iowa 50309, 210 Walnut Street
Omaha	Nebraska 68102, Nineteenth and Farnam Streets
St. Louis	Missouri 63101, Mercantile Tower, Suite 2500
Wichita	Kansas 67202, 110 East Waterman Street
Denver	Colorado 80202, 721 19th Street, Room 407
Casper	Wyoming 82601, 100 East B Street, Room 4001
Fargo	North Dakota 58102, 653 2nd Avenue, North, Room 218
Helena	Montana 59601, 613 Helena Avenue, P.O. Box 1690
Salt Lake City	Utah 84138, 125 South State Street, Room 2237
Rapid City	South Dakota 57701, 515 9th Street
Sioux Falls	South Dakota 57102, 8th and Main Avenue
San Francisco	California 94105, 211 Main Street
Fresno	California 93721, 1130 O Street, Room 4015
Sacramento	California 95825, 2800 Cottage Way
Honolulu	Hawaii 96813, 1149 Bethel Street, Room 402
Agana	Guam 96910, Ada Plaza Center Building, P.O. Box 927
Los Angeles	California 90071, 350 South Figueroa Street
Las Vegas	Nevada 89101, 301 East Stewart
Reno	Nevada 89504, 300 Booth Street
Phoenix	Arizona 85004, 112 North Central Avenue
San Diego	California 92188, 880 Front Street
Seattle	Washington 98174, 915 Second Avenue

Anchorage	Alaska 99501, 1016 West Sixth Avenue, Suite 200
Fairbanks	Alaska 99701, 501½ Second Avenue
Boise	Idaho 83701, 216 North 8th Street, P.O. Box 2618
Portland	Oregon 97204, 1220 South West Third Avenue
Spokane	Washington 99120, Courthouse Bldg., Room 651, P.O. Box 2167

INVENTOR'S GLOSSARY

Words or phrases in *italics* are also defined in this Glossary and should be referred to for fuller explanation.

ABANDONMENT. Failure on the part of an inventor to develop an invention from the idea, through continuous refinements, to ultimately filing an application for a patent.

AGGREGATION. A combination of two or more separate elements which do not perform any special or unique service, or meet any special or unique purpose by being combined. A lead pencil with an eraser at one end is considered an aggregation. Aggregations are not patentable.

ANNUAL INDEXES OF INVENTORS AND ASSIGNEES. An annual index alphabetically listing inventors' names and the names of their assignees at the time the patent is issued. Beginning with the 1955 volume, the *Index* also carries a list of patent numbers issued during the period covered by the index. These are arranged sequentially by *class* and subclass numbers. You may be able to find the *Annual Index* in a public or university library near you.

APPEAL. Objecting to a ruling in a patent case and seeking a ruling from some higher authority.

ALLOWANCE. The Patent and Trademark Office term indicating that an invention will be granted or "allowed" a patent.

APPLICATION, PATENT. The documents submitted to the *Patent and Trademark Office* on the basis of which the inventor hopes to be granted a patent. The application consists of the *oath* or declaration, the *petition,* the filing fee, a *specification* containing a description of the invention, and a presentation of the *claims.*

ART.   In the language of patents, art refers to the particular area or field in which a patent is being sought. It gives rise to such terms as *"prior art,"* "the state of the art," etc.

ASSIGNMENT.   Transferring the ownership of a patent to another individual or company. Assignments may be made in full or in part.

BASIC PATENT.   A patent covering an invention so broad that it incorporates a large portion of the "art" or technology in which it applies.

BOARD OF PATENT INTERFERENCES.   A unit of the *Patent and Trademark Office* whose responsibility is to take testimony when more than one party claims the right to an invention, and to determine which party is the original inventor. (See *Interference.*)

CHEMICAL PATENT.   A patent covering a new method of producing a chemical or a patent covering an entirely new chemical.

CIRCUMVENTION.   Getting around an existing patent by attacking its weakest *claims.*

CLAIMS, CLAIMS OF NOVELTY.   The portion of the *patent application* spelling out the aspects of the invention that make it new and unique.

CLASS.   One of the broad categories into which previously issued patents are filed. Within each class there are a number of subclasses. There are over 300 classes and more than 90,000 subclasses.

CLASSIFICATION BULLETINS.   Bulletins issued from time to time by the *Patent and Trademark Office* to supplement the *Manual of Classification* with additional information and definitions to make patent searching more accurate.

COMBINATION PATENT.   A patent covering an invention that calls for two or more units to achieve a given result. To qualify for a combination patent, the invention must produce a single result. If it produces more than one result, the invention will probably be considered an *aggregation* and the patent will be denied.

COMPOSITION OF MATTER PATENT.   A patent covering the invention of an entirely new substance.

COPYRIGHT.   The protection provided by the United States Government against copying works created by composers, writers, artists, film makers, and others. The Patent and Trademark Office has no involvement with or jurisdiction over copyrights.

DECLARATION. Part of the patent *application* in which the applicant declares that, to the best of his knowledge and belief, he is the original inventor and that his invention meets the requirements for a patent as set forth by law.

DESIGN PATENT. A patent issued for a unique design of a manufactured item.

DESIGN PATENT CLASSIFICATION. Design patents are covered by ninety-three main classes and several hundred subclasses.

DOMINATING PATENT. A patent for an invention that so completely and fully covers the field that it is virtually impossible to devise any further improvements that are patentable.

DOMINATION. Term for a patent *claim* that covers every aspect of an invention.

DRAWINGS. Part of the patent *application*; prepared in accordance with PTO standards.

FILING FEE. The fee payable to the Patent and Trademark Office for filing an *application*.

FILE WRAPPER SEARCH. A search of all of the correspondence between the Patent and Trademark Office and an inventor. When the patent is issued, this file becomes available to the general public.

FORTIFYING PATENT. A patent obtained to strengthen a patent previously issued or to cover improvements and refinements in a patent previously issued.

INFRINGEMENT. A violation of the rights of an inventor or assignee who holds a patent.

IMPROVEMENT PATENT. Any patent that is not a *pioneer patent* or a *basic patent* is an improvement patent in that it improves the "art."

INTERFERENCE. When one patent *application* comes into conflict with another because of *claims* to the identical invention.

INVOKING AN INTERFERENCE. Sometimes done by deliberately submitting a patent *application* that has *claims* identical to those of an already issued patent, so as to make an open contest over who the actual owner and first inventor of the invention really is.

LETTERS PATENT.   A document granting an exclusive right or privilege, usually to an invention.

LICENSING.   Granting another party or company the right to use your patent, usually in exchange for some financial consideration.

MANUAL OF CLASSIFICATION.   A schedule that provides a breakdown of the *classes* and subclasses, and their titles, into which patents are classified.

MECHANICAL PATENT.   A patent for a new mechanical device or for a unique combination of various mechanical components. A mechanical patent covers the article itself and not the manner in which it was produced. (*Cf. Process Patent.*)

MODEL.   A working prototype of an invention.

OATH.   A statement incorporated in the patent *application* in which the applicant swears that he believes he is the original inventor of the subject of the application.

OFFICIAL GAZETTE.   The Patent and Trademark Office's weekly publication giving the basic information on all patents issued during the week of publication. It is the PTO's official journal.

O.M.P.I.   *Organisation Mondial de la Propriété Intellectuelle.* See *W.I.P.O.*

PATENT.   The exclusive right, as specified by law, which the United States Government gives to an inventor to exclude anyone else from manufacturing, using, or marketing the invention during the life of the patent (seventeen years).

PATENT AGENT.   A registered practitioner authorized to practice before the Patent and Trademark Office and "prosecute" an invention. He is not an attorney and cannot practice in any court of law. (*Cf. Patent Attorney*)

PATENT APPLICATION.   See *Application, Patent.*

PATENT ATTORNEY.   A lawyer specializing in patents and matters related to them. Patent attorneys who are registered are permitted to practice before the Patent and Trademark Office and can represent their clients in court. (*Cf. Patent Agent.*)

PATENT PENDING; PATENT APPLIED FOR. Indications that a patent *application* is on file with the Patent and Trademark Office and intended to discourage others from trying to apply for a patent on the same invention.

PATENT AND TRADEMARK OFFICE. The government agency that, under the jurisdiction of the U.S. Department of Commerce, has authority in this country to grant patents.

PETITION. The portion of the patent *application* used to formally request the Commissioner of Patents and Trademarks to grant a patent on the invention described in the application.

PIONEER PATENT. A patent issued for an invention when the technology is relatively new.

PLANT PATENT. A patent covering a new variety of plant that has been asexually produced (except for tuber-propagated plants).

POWER OF ATTORNEY. The granting of authority to another individual to act on your behalf. Power of Attorney sometimes makes it easier for a lawyer or agent to prosecute a patent *application*.

PRIOR ART. The information that exists in already issued patents and in general literature about an invention or its elements and aspects.

PROCESS PATENT. A patent covering the process or technique by which a product is produced. (*Cf. Mechanical Patent.*)

PROSECUTION. The various legal steps required to obtain a patent.

PTO. *United States Patent and Trademark Office.*

PUBLIC SEARCH ROOM. The Search Room at the headquarters of the Patent and Trademark Office in Arlington, Va. Every U.S. patent issued can be found in the Public Search Room. Patents are grouped according to the *classes* and subclasses into which they fall.

REDUCTION TO PRACTICE. Developing the invention to the point at which it actually functions as intended.

REFERENCE TO DRAWINGS. Descriptions of *drawings* accompanying the patent *Application*.

REGISTERED PRACTITIONER.   A *Patent Agent* or a *Patent Attorney* registered to practice before the Patent and Trademark Office.

REJECTION.   Refusal by the Patent and Trademark Office to issue a patent on an invention. Rejections may be appealed.

SEARCH.   An examination of patents and publications filed with the Patent and Trademark Office in order to determine whether an invention is, in fact, new and patentable.

SEARCH ROOM, PUBLIC.   See Public Search Room.

SPECIFICATIONS.   A section of the patent *Application* that sets forth various aspects and elements of the invention and presents the *claims*.

STATUTORY BAR.   A reason for not granting a patent based on nonconformance with the statutes covering the issuance of the patent.

SUBCLASS.   See *Class*.

SUBCLASS SUBSCRIPTION.   The Patent and Trademark Office will sent to anyone copies of all future patents, as issued, in any of the subclasses he specifies. This service must be prepaid with a deposit and a service charge. The Patent and Trademark Office can supply additional details.

SUMMARY OF THE INVENTION.   A summary of the general nature of the invention, *prior art,* invention's objectives, etc. (Part of patent *application.*)

TRADEMARK.   A word, name, symbol, or device, or any combination of these, adopted and used by a manufacturer or merchant to identify goods and distinguish them from those manufactured or sold by others.

UNDERLYING PATENTS.   Those *pioneer* and *basic patents* that precede *improvement patents*.

UNITARY INVENTION PATENT.   A patent covering an invention, which involves only one unit or object. (*Cf. Combination Patent.*)

UTILITY PATENT.   A patent covering an invention of a novel and useful device, process, or composition of matter.

W.I.P.O.   World Intellectual Property Organization. An international organization of government patenting agencies whose purpose is to simplify the filing of patent applications among member nations.

INDEX

81
8